Science and Fiction

For further volumes:
http://www.springer.com/series/11657

Science and Fiction – A Springer Series

This collection of entertaining and thought-provoking books will appeal equally to science buffs, scientists and science-fiction fans. It was born out of the recognition that scientific discovery and the creation of plausible fictional scenarios are often two sides of the same coin. Each relies on an understanding of the way the world works, coupled with the imaginative ability to invent new or alternative explanations—and even other worlds. Authored by practicing scientists as well as writers of hard science fiction, these books explore and exploit the borderlands between accepted science and its fictional counterpart. Uncovering mutual influences, promoting fruitful interaction, narrating and analyzing fictional scenarios, together they serve as a reaction vessel for inspired new ideas in science, technology, and beyond.

Whether fiction, fact, or forever undecidable: the Springer Series "Science and Fiction" intends to go where no one has gone before!

Its largely non-technical books take several different approaches. Journey with their authors as they

* Indulge in science speculation—describing intriguing, plausible yet unproven ideas;
* Exploit science fiction for educational purposes and as a means of promoting critical thinking;
* Explore the interplay of science and science fiction—throughout the history of the genre and looking ahead;
* Delve into related topics including, but not limited to: science as a creative process, the limits of science, interplay of literature and knowledge;
* Tell fictional short stories built around well-defined scientific ideas, with a supplement summarizing the science underlying the plot.

Readers can look forward to a broad range of topics, as intriguing as they are important. Here just a few by way of illustration:

* Time travel, superluminal travel, wormholes, teleportation
* Extraterrestrial intelligence and alien civilizations
* Artificial intelligence, planetary brains, the universe as a computer, simulated worlds
* Non-anthropocentric viewpoints
* Synthetic biology, genetic engineering, developing nanotechnologies
* Eco/infrastructure/meteorite-impact disaster scenarios
* Future scenarios, transhumanism, posthumanism, intelligence explosion
* Virtual worlds, cyberspace dramas
* Consciousness and mind manipulation

Brad Aiken

Small Doses of the Future

A Collection of Medical Science Fiction Stories

 Springer

Brad Aiken
Miami
USA

Locked In — First printed in *Analog Science Fiction and Fact* in March, 2010
Questioning the Tree — First printed in *Analog Science Fiction and Fact* in July/August, 2010
Freudian Slipstream — First printed in *Analog Science Fiction and Fact* in December, 2011
Done That, Never Been There — First printed in *Analog Science Fiction and Fact* in September, 2012
Hiding from Nobel — First printed in *Analog Science Fiction and Fact* in December, 2011
The Last Clone — First printed in Analog *Science Fiction and Fact* in April, 2013

ISSN 2197-1188   ISSN 2197-1196 (electronic)
ISBN 978-3-319-04252-7   ISBN 978-3-319-04253-4 (eBook)
DOI 10.1007/978-3-319-04253-4
Springer Cham Heidelberg New York Dordrecht London

Library of Congress Control Number: 2014932667

Printed on acid-free paper

Springer is part of Springer Science+Business Media (www.springer.com)

Foreword: Ten Random Facts About Brad Aiken

FACT THE FIRST: When Brad Aiken was just an infant the Grand Vizier of his repressive desert nation was so concerned about the prophecy that a hero would rise that he sent the army forth to slay every first-born male child. His mother placed him in a wicker basket and sent him floating down the river, in the hope that some wealthy person would find him and raise the babe as his own. He was found by a woman who didn't mind getting the baby but thought the basket was just the perfect thing to fill with flowers and set on display in the breakfast nook. The baby was just an extra.

FACT THE SECOND: Brad Aiken makes his living as the head of Rehabilitation Medicine at Baptist Hospital in Southern Miami. He is a compassionate and assured doctor whose professionalism is a substantial note of light and encouragement among those whose physical ailments bring them under his care. He carries his gentle professional demeanor into his personal life, to the point where this friend has never, repeat, never, heard him raise his voice in anger.

FACT THE THIRD: In his twenties, Brad Aiken drifted to the flesh markets of Marrakech, where he swiftly organized a network of thieves and assassins who would come to be feared on eight continents—more, you'll note, than there actually are, which is just another measure of just how respected they were. It was said that the men and women under his employ could hide in shadows, break into the most secure vaults, and steal a fortune in jewels without anyone ever suspecting that they were there. It was also said that they got together on weekends for paddle-ball.

FACT THE FOURTH: Brad's love of science fiction is visible in his home. He has big colorful paintings of the starship Enterprise soaring majestically over ruined worlds, and a book and DVD collection that favors the colorful and fantastic. He's a professional now, but if you want to see this eminently grown-up man become twelve, you just need to watch him at a science fiction convention. His eyes become wide. He is impressed with every accomplished writer he meets and does not particularly seem to believe that he can now be counted among their ranks. The man has little self-aggrandizing ego. He has a wholly unironic wonder that shows in his stories.

FACT THE FIFTH: Brad built a raft out of balsa wood and attempted to use it for a solo Pacific crossing. Leaving from Los Angeles, he got within about a foot of Africa when he realized he had run out of supplies, and went back. The brass band that had gathered to celebrate his arrival was, of course, bemused.

FACT THE SIXTH: Brad and his lovely wife Laura are the proud parents of two fine adults, Danielle and Dustin, who right now sit at the entrance to all the possibilities that life has to offer. You want to intuit everything you need to know about Brad and Laura, you look at them.

FACT THE SEVENTH: The many years Brad spent as a prisoner in solitary confinement in the repressive European nation in Slobistan were years of tremendous cultural and political upheaval which he, of course, missed completely, because he was a prisoner, in case you weren't paying attention. During his sentence he learned how to play the harmonica and made sure that everybody else on his cell block understood that nobody knew the troubles he'd seen. This is precisely the kind of thing that will make other prisoners feel excluded, since they're in cells too. This was a rare failure of empathy on Brad's part.

FACT THE EIGHTH: Brad is a member of the South Florida Science Fiction Society's writing workshop, which saw early versions of stories like "Questioning The Tree." He takes to criticism about as gently as any writer I have ever seen; when a point is scored, he simply smiles and chuckles softly and applies it to his own aesthetic judgment, without ego. (Bonus fact: he happens to have one of the most likeable laughs of anybody I know. It's usually just a gentle little heh-heh-heh, but if you can get him laughing uncontrollably it's one of the great pleasures of life; he gives himself over to it fully, and it's impossible not to love him a little bit more when you hear it.)

FACT THE NINTH: Brad spent his thirties combing the world for the mysterious nobleman he only knew by the ampersand-shaped scar on his right wrist. Toward this end, he fought off assassins, seduced beautiful women, stowed away in secret military transports, crossed heavily guarded national borders, wrestled bears, clipped coupons, and finally crawled through three hundred meters of air shafts to drop down into the secret chamber where the Duke himself sat plotting world domination with his faithful minions. "You!" cried the Duke, who had been following Brad's attempt to find him, for years. "What do you want?" Brad reached into his pocket and dropped something on his table. Seems the Duke had left his Visa Card at the restaurant in Delray, and could have had it back from Brad a lot sooner if he hadn't been such a turd about it.

FACT THE TENTH: Brad is still, by the usual measurement of such things, a new figure on the science-fiction scene. His association with ANA-

LOG magazine, profitable for both the venue and the writer, showcases his skill at capturing the ethical dilemmas that come along with new technologies, and the tough choices that need to be made by the human beings in the way of change. He is still at the point where he is discovering, story by story, just how good he is capable of being. I think this level of surprise will continue for quite some time. When you complete this book, you will still not have seen the best of Brad Aiken. That's still coming. But what's here is pretty damn good, and will be a treasured experience for you. I envy you, first-time reader, your first encounter. Ladies and gentlemen, here's Brad.

Adam-Troy Castro

Preface

One of the great debates in science fiction these days is whether it is losing its relevance as the rate of technological advancement increases. The reality of new technologies often outpaces that which has been predicted in literature, and nowhere is this more apparent than in Medicine.

Forty years ago CAT (Computerized Axial Tomography) scanners revolutionized the way we image the body, and in the decades since there has been a rapid-fire release of one technology after the next—MRI (Magnetic Resonance Imaging), CTA (Computerized Tomography Angiography), MRA (Magnetic Resonance Angiography), PET-CT (Positron Emission Tomography—Computerized Tomography), etc.—that allows us to see inside the body at a level of detail that was considered impossible only a few short decades ago. In fact, today's imaging rivals that depicted in early Star Trek episodes that were trying to predict the technology of the twenty-third century.

Advances in robotic surgery have improved the accuracy of skilled surgeons; new medications utilizing nanotechnology are improving efficacy while diminishing side effects; electronic health records are rapidly disseminating information to all involved providers; and an ever-increasing array of new technologies has contributed to the evolution of modern Medicine.

In my field, Rehabilitation Medicine, technologies such as brain-computer interfacing (allowing someone to directly control a computer with only their thoughts), bionic prosthetic limbs, robotic therapy devices (robots that aide in physical therapy), and robotic exoskeletal braces that allow paraplegics to walk again have already crept out of the pages of science fiction and into the real world. Although most of these are not yet advanced enough for widespread clinical use, there is no doubt that they will dramatically improve the way we help people with disabilities regain their independence.

The stories in this collection look at where we might be going in the near and distant future regarding not only the continued application of these remarkable technologies, but also the political and ethical implications of how all this technology is going to affect us as individuals and as a society. And while it is fun to look at how the technological advances in medical care may enhance our lives, the ethical dilemmas posed by these changes are perhaps even more intriguing.

I hope you'll find these stories not only entertaining, but also thought-provoking. After all, the future is not set in stone, and it is our job, all of us, to try and nurture it to grow in the right direction.

Many people have helped me to create the stories in this collection, the most important being my wife, Laura, who shares my fascination with scientific postulation and discovery, and has encouraged me to forge on with my passion to write. I was incredibly fortunate to cross paths with Stanley Schmidt, the Hugo Award winning editor of *Analog Science Fiction and Fact*, before his retirement earlier this year. I learned a great deal about crafting a story to make it thought-provoking and true to the science behind the story by attending his Worldcon panels, reading his editorials in *Analog*, learning from his comments about my stories (including a stack of those he rejected, which often taught me the most), and his meticulous editing.

There's nothing like a workshop full of friendly faces looking to rip your stories down to the core (in a good way!) with their no-holds-barred critiques—an invaluable resource. I have been blessed to be in a workshop run by Adam-Troy Castro and with input from Ben Burgess, Judi Castro, Dave Dunn, Cliff Dunbar, Chris Negelein, Giselle Peterson, Mark Halverson, and David Slavin. I can't thank them enough.

One final nod to Dr. Nick Kanas, who I had the good fortune to meet at Worldcon this year. Nick informed me about the *Springer Science and Fiction* series just as I was looking for the right fit for this book. He connected me with Dr. Christian Caron, the editor of Springer's series, and the two of them helped me organize the 'science and fiction' thoughts behind this collection. Without them, this book would not exist.

Brad Aiken   November 2013

Contents

Part I

The Short Stories 1

1   Locked In 3

2   Questioning the Tree 17

3   Freudian Slipstream 33

4   Done That, Never Been There 45

5   Hiding from Nobel 95

6   The Last Clone 113

7   If He Only Had a Brain 125

8   Once, on a Blue Moon 143

9   And a Time to Every Purpose 161

Part II

The Science Behind the Fiction 179

10   The Invasion of Modern Medicine by Science Fiction 181

     References and Suggested Readings 191

Part I

The Short Stories

1

Locked In

This one is near and dear to my heart, as it was my first story published in Analog. It was also inspired by one of my first patients, an executive who suffered a devastating kind stroke referred to as Locked-In Syndrome, where the brain is essentially disconnected from the rest of the body, allowing the victim to think normally while denying its victim the ability to move any part of their body, to speak even a single word.

Upcoming technologies such as brain-computer interfaces and robotic exoskeletal devices hold great promise in giving someone like this the opportunity regain control of their life, but the process of recovery is still going to be an arduous one that is likely to change anyone going through it.

Sometimes we have to give up a lot to gain a new perspective on life, and sometimes it makes us a better person. I doubt many would choose this trade off, though.

At the time, I didn't know it would be the last few seconds of my life, at least what most of us would call life.

I eased back into the oversized black leather chair that marked my spot at the head of the boardroom, listening to the same old inane arguments about why the privileged few of us sitting around that table should have had the right to define ethics a little differently than everyone else; why we *needed* to define it differently.

Bullshit. Those new guys may have been Ivy Leaguers, but they didn't know squat. Maybe I wasn't so different when I started out, but meeting Linda changed all that. "Never make a decision you can't sleep with," she'd told me, once. "Then go out and do it better than everyone else."

I wish she was still here to see how right she was.

Like most businesses, this one's always been populated by ass-kissers, and over the years it hasn't been too hard to get people to do things my way. But that new generation was different. Dave Dunnster had taken over old man Reiss's spot almost 3 years earlier and got under my skin right from the start.

B. Aiken, *Small Doses of the Future,* Science and Fiction,
DOI 10.1007/978-3-319-04253-4_1, © Springer International Publishing Switzerland 2014

The younger board members may have respected me, but they could identify with Dave and he knew how to work a room.

I decided to let him have his say before reminding everyone that I hadn't built this company into a Fortune 750 powerhouse by accident. Even in this business there are some things you can't compromise on; *especially* in this business. I couldn't believe they were stupid enough to not see that.

Tuning out the noise, I looked past the blithering idiots sitting around the table to the view out the 103rd floor picture window that had sold me on that office 28 years ago. A small etch mark in the lower right side of the middle panel, made by a careless window washer who couldn't balance his platform, had been growing like a spider web; I made a mental note to have Gladys call the maintenance crew chief again.

"For God sakes, Troy, how can you let that slide?" The familiar voice pulling me back into the conversation was that of my best friend, Tanner Hopkins, who had co-founded Nucleic Innovations with me back in 2015. "Are we really going to be part of this madness?"

I looked him squarely in the eye. *Over my dead body!*

"Troy?"

The twelve men around the table all gawked at my silence.

What the hell are you all staring at?

It was at that moment I realized the words had never left my mouth. I grabbed the blue Waterman pen out of the inside pocket of my blazer and scribbled as quickly as I could on the pad in front of me, which was embossed with our company's insignia, the classic image of two electrons orbiting a neutron superimposed on a vertically oriented missile.

Before I made it through the first sentence, I could feel the strength draining from my hand. I wrote faster, but the markings soon became illegible. The pen dropped to the table and I slumped forward, helpless to avoid the painful crack of my skull against the wood.

"Jesus, Troy," Tanner said. "You OK?"

I couldn't answer, couldn't even figure out a way to show him I was still alive, short of hoping he'd notice my eyes blinking.

"Troy!"

They tried to pull me up, but I started to slide under the table like a lump of Jello, unable to let loose the screams that were ricocheting around inside my head. It took a group effort to get me back to the same awkward position I'd started in.

Since the moment that pen had dropped from my hand, every fiber of my being had been concentrating on regaining control. And then I saw the look on Tanner's face.

I was going to die.

The wave of terror welled to a deafening roar.

Tanner disappeared from view; the unmistakable metallic sound of the handle to the fire-door separating us from the outer office was followed by a symphony of sounds pouring in from the outer office.

"Somebody call 911!"

My head bobbed back and forth against the glossy wood as someone tried to shake me back to life. The office buzzed with murmurs and the sound of chairs scooting across the floor, each one reverberating like a Saturday morning alarm clock you forgot to turn off the night before.

"Christ, he looks bad."

"Hang in there, Troy," Tanner said. Then, the other direction, "For God's sake, hurry. It's Mr. Adams."

God knows how many of them crowded around the table. The voices, asking each other if I was moving, if I was dead. All I could see was the damn table top a half inch from my eyes, but I felt the pressure of dozens of gawking faces hovering just beyond my field of vision. I never thought I'd be so happy to be whisked off by paramedics.

The blare of sirens droned above my head and the white steel roof that filled my field of vision flickered with each bump in the road. The EMTs poked and prodded, blinded me with their flashlights and bombarded me with questions I could only answer with silence. I looked back and forth as the two of them took turns grilling me. When they finally stopped, I closed my eyes tightly and took solace in the darkness.

If only Linda were still alive.

I tried to keep my eyes shut as the gurney was jostled across the pitted sidewalk into the ER, where we were stopped by a gravelly voice. "What do you got, Harve?"

The now familiar voice of one of my couriers sputtered back. "Not sure. Vitals are stable as a rock and he's looking around, but I can't get him to do a damn thing but blink every now and then. I haven't seen one of these in a while, but it looks like he's locked in."

*

'Locked in.' Now there's an appropriate term for it. I surely was. It was the first time I'd ever heard that cursed diagnosis, and after ten more hours of tests the neurologist on call confirmed what the paramedic had guessed in 10 minutes of scrutinizing my now worthless body.

I was locked inside a motionless shell. I could hear and smell and see, I could think as clearly as ever, I could feel the itch of my nose. Every fetid breath of every person who leaned over to examine me burned my nostrils with an intensity I'd never known before; there wasn't much else to entertain my senses. My head floated, detached from a lifeless body, a floppy mass of muscles that had no one telling them what to do.

I lay on a stretcher, unable to avert my gaze from the brilliant light of the procedure suite in the vascular institute, and listened as the interventional radiologist explained the procedure to his resident in vivid detail. My skin was now foreign to me, but I could feel the catheter as it was threaded up through the femoral artery, aorta and then into the basilar artery, that tiny little bastard that cut me off from the world.

"There. Got it!" Dr. Sitieu gloated. "Hooked it like the cork on a bottle of fine Merlot."

The radiology resident was impressed. "Man, look of the size of that sucker."

"Second biggest I've ever seen," Siteu said with pride as he pulled the clot from the artery at the base of my brain, threaded it back out through my body and deposited it on the instrument tray next to the table. I caught a glimpse of it on the LCD monitor to the right of the procedure table.

Siteu saw me straining to look and turned my head to help the cause. "So what do you think of that?"

It looked like a juicy purple earthworm.

"That's the sucker that was clogging up the works," he said. "Feel any different?"

One blink for yes, two for no, they had told me. I blinked twice.

"It may take a while. Let's see how we did."

I felt a warm flush as he injected dye into the artery he'd just pulled the clot from. A streak of white puffed out of the tip of the catheter on the wall monitor. I wasn't sure what it meant, but the doctor seemed very pleased with himself.

"Perfect." He turned to leave the room.

All that night I lay in the hospital bed staring at the ceiling. Dr. Siteu's words were encouraging, but I couldn't feel a damn thing. The procedure was a success, but I was not.

*

Every time I closed my eyes, I saw her face. It was an image I had tried so hard to erase, but never with more than fleeting success. If I'd only left the party when she pleaded with me, if only for once in my life I had let business take a back seat to what I couldn't see was the most important thing in my world, maybe I wouldn't have been so tired, maybe I would have seen it coming in time to stop.

I had regained consciousness just long enough to see her sitting beside me, that beautiful face all bloodied, just inches from my own as she took one last breath before I lapsed back into coma knowing it would be the last breath I'd ever see her take. That look has haunted me for almost 11 years.

Morning was a welcome relief. Nurses and doctors started filtering in. Hell, I was even glad to have my blood drawn. But it wasn't until the neurologist gave me the once over that I realized I could move my right thumb an eighth of an inch; the procedure had indeed been a rousing success.

And so I lay there twitching my thumb. The day was interspersed with visits from therapists and brief depressing hellos from a parade of business associates who felt obliged to pay their respects, but could rarely stomach looking at me for more than a few seconds. Tanner would sit and bring me up to speed on the goings on at the office, but it was hard for him; he didn't know how much was getting through to me, and even *he* showed the restlessness of a discomfited sick-bed visitor.

The days were interminable.

And then came the nights.

And Linda.

I'd learned to live with the pain, to banish the terrors by filling my head with other images, any images besides that battered lifeless face. She was the best thing that ever happened to me and all I wanted was to forget her.

But now there was nothing else. No conversation, no books, no mindless escorts to stimulate my senses into pleasure divorced from joy. I couldn't even turn on the damn boob tube. Even in the worst of times, there had always been the promise of tomorrow, a hectic office, a roost to rule.

Tomorrow…well, work was out of the equation now. There would only be more time to think.

And in between those unbearable moments came a parade of physical therapists, or as I came to know them, physical terrorists, who would take turns marching into my room every morning and again in the afternoon, stretching every joint in my body until tears ran from my eyes. It was their only signal that I could take no more. They tried to convince me that it was critical to maintain my range of motion to keep me from turning into a pretzel that would be in constant agony and begin to rot at every convoluted crevice that would harbor bacteria and fungi; a delightful picture that made the torture of therapy no easier to bear.

It was 2 weeks to the day after my stroke that the rehab doctor gave me the first glimmer of hope.

"Mr. Adams, we haven't gotten too far with your therapy."

Thanks for the insight, I said inside my head as I lay motionless in bed. *I never could have figured that out without you. Is that what 8 years of med school and residency get you?* I looked away and tried to tune him out. *Asshole.*

"But there is another option."

Wait. *What?* I looked into the eyes of Dr. Benson Burgess. Empathy. There's been a lot of that since I landed on the Rehab Unit, but I need more than that this time.

"A computer-brain interface."

I stared blankly.

"Have you heard of it?"

I blinked twice.

"It was developed around the turn of the century at Duke. First on monkeys, then later on humans with spinal cord injuries. It's a way of getting your brain to send signals to a computer that's programmed to control a robotic arm, a powered wheelchair, even a voice generator. It's not easy to master, but results have been pretty good in some cases."

Great. I'm a case *now. What the hell; at least it's a chance.*

"It takes a lot of money and a lot of work. It also involves cutting your head open and implanting a series of microchips on the surface of your brain. We haven't seen many complications, but it could cause a brain infection, seizures, even death if things go wrong. You still interested?"

Like I've got a lot of options. I blinked once.

The surgery took place 1 day later. I awakened in my room feeling no different than I had before the procedure; couldn't even get a look at myself or grope my skull.

Dr. Burgess entered with a wide grin. "It went great," he said. "Wanna see?"

I blinked once.

He held up a mirror he'd carried in with him. The left side of my head was shaved and had a u-shaped incision that was stapled shut. I looked like a baseball with hair growing out one side.

"Your hair will cover all that in a few weeks," he said. "The neurosurgeon placed seven chips along your pre-central motor strip."

My face was obviously blank, but somehow he caught my confusion.

"Uh…the surface of your brain; the part that controls your right arm, your right foot and your speech. The chips will pick up the brain activity, send it through the wire he tunneled down to you belly." He showed me the faint bulge under the skin on my neck and chest, and the small incision in my abdomen. "…and trigger the IR transmitter here." He pointed to the incision. You'll wear a belt that will pick up the IR impulses and turn them into RF waves that are then sent to the computer, which will turn your thoughts into actions."

I hadn't thought it possible, but yes, the grin had gotten wider.

"Got it?"

I blinked once. At least I got the gist of it.

"Good. The first goal is to get you talking. Once we do that, the rest will go a lot faster." He positioned the computer monitor in front of my face. "Now say hi."

Hi.

Nothing. So far, I was not impressed.

"No, no. Don't *think* hi, *say* hi. You've got to use the same part of your brain you would if you were talking. I know you're out of practice, but it's like riding a bike… OK, bad example, but you get my drift."

I did. *Hi,* I said again, and the voice came out of the computer almost as I said it. *Shit, it sounds just like me.* That last part just came out as garbled nothingness.

Burgess grinned. "Pretty cool, huh? The speech therapist programmed it with recordings we had from your office. It sounds just like you."

He understood me?

"Well, not that last part; that sounded like mush. The computer's got to learn what the signals coming from your brain mean. I programmed it to say hi in response to the first signal you sent, so hopefully you weren't screwing with me, because whatever you said first is going to come out as hi every time you say it from now on. Was it hi?"

I blinked once.

"Good. We'll do yes and no next, then if all goes well, I'll have the speech therapist get started with some vocabulary."

I spent the better part of the next month with a brunette named Marta teaching me how to talk through the voice synthesizer. She was deceptively demure, unrelentingly tenacious, and worked me to exhaustion; the verbal equivalent to my physical terrorists. One by one, she flashed pictures up on the monitor, I named the objects I saw, and the computer learned which signals from my brain corresponded to which words. Tedious, but very cool. I grew to look forward to these sessions. No doubt I would soon become little more than a memory to Marta, another patient in her files, but her daily visits came to bring a familiarity that I treasured in this chaotic abyss I'd fallen into.

And when I was not in therapy, my mind had time to play with my sanity. The harder I tried to forget, the more I remembered.

But in forcing me to confront all I had suppressed for the better part of a decade, my mind began to let me remember the woman I had loved. Memories of all the years before the accident began to wash over me. Our wedding, the briefcase she surprised me with as I left for my first day on the job, vacations and endless conversations about life, about kids. We both wanted them, the time was just never quite right. And then one day there was no more time.

So alone.

*

Burgess had just returned from a trip to Israel to meet with some robotic engineers who were working on a new type of wheelchair for me. "So," he said, "how's my favorite patient?"

"Tired of lying in this damn bed staring at the ceiling."

He smiled. "I see Marta's gotten around to the practical stuff. Good."

"Sweet kid, Marta, but enough's enough. When are you going to get me out of here?"

"Funny you should ask."

He walked out and a few seconds later, there was a thump as the door burst open and The Monster—his words, not mine—whirred into the room with Dr. Burgess close behind. It was a wheelchair with a big black metal box on the back, which I assumed held the battery and computer, and a robotic arm tucked in along the right side.

"My new jail cell?"

"Your freedom. You're going to learn how to maneuver the chair with your thoughts and control this arm as if it were your own."

My own…it looked like the detached arm from the terminator robot in the 3D remake of that old Schwarzenegger flick, a shiny mass of metal and cables in the form of a human limb without the skin. Only instead of being attached to a torso, it protruded from the end of a telescoping titanium rod emerging from the right side of the chair, where it pivoted on a ball joint of a shoulder. The dexterous fingers were going through a demo-mode that made it look like they were the metallic equivalent of a disembodied magician's hand practicing the motions of a well-ingrained card trick.

I wasn't sure whether to be impressed or to vomit.

"Will I be able to play the piano?"

He laughed. "Ah, the age old joke. Glad to see you've got your sense of humor back."

"I was a concert pianist before I went back to business school," I said.

"Oh…I, uh… sorry." Burgess turned beet red.

It felt good to be able to make someone squirm again. "So let's get to it. I'm going buggy here."

The next month was almost as tedious as the last. I was assigned a new team of therapists to show me how to use all the gadgets I'd need to function at home. My latest physical therapist, a brute of a guy named Bill, turned out to be a lot less intimidating than he looked. He taught me how to activate the automated transfer sequence that slid my body from the bed into the chair and back. Once I mastered that, the rest seemed easy by comparison, almost intuitive.

I learned a lot more than I ever wanted to know about being disabled, and even more about the recovery process. In spite of all the cool technology, it took a crapload of work to accomplish even the most mundane things I'd always taken for granted; feeding myself, brushing my hair, that kind of stuff. Other things like getting dressed, washing, even wiping myself…forget it, the robots aren't that advanced yet. Not sure I'd really trust a robot to wipe my butt anyway. Your definition of vanity changes, to say the least.

An occupational therapist, as it turns out, has nothing to do with occupations of any kind, except their own of course. It was Laura Feibelman's task to make me more independent in activities of daily living, which in my case boiled down to figuring out how to use the robotic arm. All I had to do was think about moving my own arm, and the thing mimicked my thoughts – not as easy as it might sound. Manipulating a soupspoon with a 73 pound hunk of machinery strong enough to bend metal can be a little nerve wracking. But practice makes perfect, and what else did I have to do?

The Monster's drive mechanism was a blast to learn. I could steer it with my new arm using a joystick, and to accelerate, all I had to do was push down on the pedal. My foot didn't actually move, of course, but my brain didn't know that that unless I was looking down; the computer did all the work, but the end result was the same. It only took a few minutes to learn how to accelerate and brake just like driving a car.

Eleven weeks to the day after my stroke I was ready to return to my world.

Going home was the hard part. I'd grown so accustomed to everyone at the Rehab Center, it felt like my training wheels were being taken off and I was being pushed into oncoming traffic. Of course, it wasn't nearly that bad. Two nurse's aides had been working with me and the rehab staff for the past week learning how to do whatever I'd need to have done at home. And the house was equipped with full environmental controls that let me manipulate everything through my computer-brain interface by tapping into the home automation system. Lights, heat, TV, telephone, and of course the pager that my aide wore on her wrist were all at my beck and call.

It's nice to have money.

Still, it didn't feel like home. Linda should have been there.

My newly appointed driver picked me up the next morning. Very tacky. I made a note to have them build me a van I could drive myself.

He unloaded me at the main entrance. It felt odd to be wheeling into the atrium at Nucleic Innovations through the same doors I'd walked through a thousand times before.

Tanner had been keeping me abreast of the boardroom battles going on in my absence. I whirred into his office in The Monster, bruising the polished wood as I entered.

Tanner started at the cacophonous intrusion.

My brain smiled, but I knew all Tanner could see was a blank face. "Guess I'll have to get some kick plates on those doors before I punch holes in them."

"Or you could wait for somebody to open it," Tanner said, surveying the damage.

I ignored him. "So what's happening with the Shower Heads?" Shower Heads…we always managed to come up with some innocuous name for each of our brutal inventions; in this case, missiles carrying a head of nuclear waste that would burst three hundred yards over its target and shower the area with its venom.

"We vote on Thursday," Tanner said. "The Defense Department's pressuring us to come up with the goods or they're going to release it to another contractor."

I tried without success to bite my lower lip. "You know I'm not one to shy away from a new weapon system. Hell, we built this place on nuclear applications for war, but up till now it's all been miniature explosives like the Pea Shooter Grenades or nucleic power cells to run the tanks and planes. But this…how will history judge us, Tanner?"

"Hell," Tanner said, "how will we judge ourselves?"

"It would certainly end the argument about whether we crossed the line in letting the ethics of war justify what we do."

"So how do we stop it? The board's split dead even on this one."

"In which case, my vote decides the victor," I said. "It's good to be king."

"I was afraid you wouldn't cash in that chip," Tanner said.

I had never had any problems impressing my will on others, winning by conversion rather than coercion. A press of the hand, a gleam in the eye, a nod so subtle it only registered in the direction it was aimed. I could sway any vote my way and everyone would walk out thinking I was a nice guy; depends how you define nice, I guess. I ruled with an iron fist that no one noticed, but life would be different now; no time for nuance.

"I've been saving it for the right occasion."

When Thursday morning came I was ready to do battle. God, it felt good to mean something again. In spite of a crowded lobby, I had the elevator to myself; The Monster can be pretty intimidating. The bell dinged and when the doors slid open I rode out with purpose, right past the gawking secretaries and through the newly installed automatic door to my office.

Tanner appeared at the doorway a few minutes later. "Ready?"

"Let's get the bastards," I said.

Tanner led the way and held the boardroom door for me as I rolled in. We were the last to arrive. My customary spot at the head of the table had been cleared to make room for the new mechanical me.

Dave Dunnster stood. "First on the agenda is the financials from the fourth…"

"We're not blind, Dave," I sputtered. The agenda was up on our monitors plain as day.

"But it's protocol, Troy."

"Screw protocol. Let's get to the meat, shall we gentlemen?"

The buzz around the table was comical. I had always placated these turkeys and it was fun watching them adjust to my change in attitude; some experiences can affect your priorities. It was tedious using my new brain toys and I had to get business done before I collapsed from exhaustion.

"Right, then," Dave said. "I assume you mean the Shower Heads."

I nodded, but my head didn't move. "Shit," I muttered, then louder, "that was me nodding yes. Everybody ready to vote?" Of course they were. It was no big secret who was voting which way.

We went around the table with alternating yeas and nays, 6:5, to me.

"I vote nay, which knots us up, and I'm invoking executive privilege to break the tie. The Shower Heads are off the table. Next order of business."

Negelein and Christopher looked stunned; they were squarely with Dunnster on this one, the prime movers to break me and send our horrific weapon to friendly governments all over the world. Negelein stood. "But you always said no one's vote should count more than anyone else's, Troy. That we're a democracy here."

"I lied. Next order of business." I looked over at Dave, who was tapping at his mobile.

"I, uh…under regulation sixteen, I request a recount."

I rolled my eyes skyward, one of the few facial expressions I was still capable of. *Who's mind you gonna change?* I said, but what came out of my synthesizer was "Of course, Dave."

I looked over at Tanner, who was staring in disbelief.

"Go ahead, Dave. Convince me." My synthesizer said. *What the hell's going on!* I screamed in silence.

As Dave started to talk, I wheeled back and to the left. Thank God I still had control of my chair. I pushed ahead 6 ft, then turned right, to face the picture window.

"Troy?" Tanner said. "You all right, Troy?"

Hell, no. I pushed the accelerator in my mind, pedal to the metal, and headed straight for the glass with my footrest aimed at the spider web crack in the lower right of the middle pane.

The Monster was fast. I struck with the full force and considerable weight of the beast, following the shards of glass as they shot out from the 103rd floor window, arcing to the street below. I watched the surreal image unfold before me, strapped securely into my chair which now dangled precariously from the end of a robotic arm clasped to the steel frame of the missing window.

"Troy!"

I could hear the simultaneous cacophony of yells behind me. *You alright?... Are you nuts?...My God, he's slipping!...What do we do?* Then a solitary *Don't fall, Troy!*

Now there was some good advice.

The lone voice of reason came from Judi Cee. "Hold on, we'll pull you in…Somebody get something…the fire hose, wrap it around the back of his chair."

They had me safely on the floor of the board room within minutes, still strapped into my chair. I wasn't about to hang around to see what else they'd have me say. I made a bee line for my office, with a trail of harried men in suits close behind.

The door slid open and I rolled in, then spun toward the pack of suits. "Tanner, stay. Everybody else, out." The synthesizer was obeying me once again.

The jumble of commotion ceased and a roomful of blank faces stared back at me.

I said, "Sooner would be better than later."

A few of them nodded respectfully, others avoided my gaze altogether as they turned and shuffled out. I couldn't tell which ones the offended grunts came from. Dunnster's stare lingered, but he soon followed, and the door slid shut behind him.

Tanner stood in front of me with his arms crossed. "What the hell was that all about? Have you totally lost it?"

"In a manner of speaking," I said. "Dunnster bugged the board room. He figured out a way to hijack my synthesizer; must have rigged up some sort of WiFi hack. Those were his words you were hearing in there, not mine."

"Well that's a relief."

"Not to me. I'm used to being the puppet-master, not the friggin marionette."

"Saw what it was like to have the shoe on the other foot, huh?"

"Damn right. It sucks."

"I suppose that kind of thing can change a man."

"I suppose."

As Tanner studied my expressionless face, a subtle grin crept across his lips. "But not you."

"*Hell* no."

"So what do we do now? If anyone finds out your synthesizer can be hacked, they'll try and pull you off the board."

"We keep it between us. I doubt Dunnster was dumb enough to let anyone else in on his plan."

"Probably right, but even if they don't give you the boot for that, they're liable to try and get rid of you for going suicidal. What the hell were you thinking in there?"

"It was the only way I could stop the vote. I couldn't just leave."

"Right. We were already underway. You'd have just counted as an abstention."

"And if I stayed, Dunnster was in control. I had to find a way to stop the vote and make sure we could move it to another venue, one where I can speak for myself."

"You think they're going to just forget what happened in there?"

"I'll tell them The Monster malfunctioned; the drive control went haywire. I've already hacked into the software and put a bug in the drive program. I'll activate it when we're done here. Just shut down my drive battery before you leave, then tell everyone what happened and call them in here for the vote. Pull in Clifford from IT; he'll confirm the malfunction, you'll pronounce me mentally fit and we'll get this done today."

"What about Dunnster?"

"He's toast. I can tap into the mainframe now; it's easy to become a computer expert when your brain is part of the computer. With a few well placed thoughts tied into Dunnster's personnel file, I'll make sure he's got his hands full just trying to stay out of jail. In fact, I think I'll tip him off before letting anyone discover his newly created transgressions just so I can watch him squirm."

"You're a scary guy, Troy."

I smiled inside. "Start preparing a short list for Dunnster's replacement."

Tanner fixed on the picture hanging on the wall behind me, the aftermath at Hiroshima. Perspective. His eyes narrowed and tracked toward my desktop monitor, then without diverting his gaze he made his way slowly around behind me to see the image that had been reflecting in the glass in front of Hiroshima, the tenth anniversary picture his wife had taken of me and Linda at the Starlight Grill.

He smiled and put a hand on my shoulder, then walked out the door without saying a word.

And for the first time in years I looked into Linda's eyes, thankful for the strength I could once again find there; more thankful yet that she would never see me like this.

2

Questioning the Tree

Here's a story that quite obviously reflects my frustration with the growing trend of trying to diminish the role of the physician in medical care in favor of technology, rather than having technology augment the capabilities of the physician.

Questioning the Tree was meant as a bit of a tongue-in-cheek caricaturization of this type of technocentric system of health care, flavored with the paranoia of a litigation system run amok. Much to my dismay, this story seems a lot less far-fetched today than it did when I wrote it.

It wasn't such an unusual sight, but it was the first time I'd seen it live, the first time they'd snared one of my own colleagues.

I had just come through the revolving doors that deposit visitors into the opulent lobby of the Metro Towers building like a Pez dispenser, from nine to five every day. Across the white Italian marble floor at the far side of the atrium, camera crews for every major news company lay in wait for something deliciously ominous; had to be, to draw this kind of attention.

An elevator door slid open in front of the throng and flood lights poured into the space revealing a middle-aged man with tightly cropped gray hair twisting away from the brightness. His hands were cuffed behind his back, each arm in the grasp of a uniformed federal officer. As he turned in my direction, the familiarity of his bushy gray eyebrows, ruddy complexion and paunchy abdominal girth sent a chill up my spine; it was Arnie Hirsch, an old friend who'd joined my practice at the District 13 Medical Clinic 7 years ago.

A familiar voice behind my right shoulder startled me. "Veered from the answer tree."

I turned toward my new assistant, Carma Johnson. "What?"

"They got Dr. Hirsch on an answer tree violation. His third one."

I knew what she meant, of course. We were only allowed to say certain things, specifically scripted responses to questions that were always variations of the same things: *What do I have, Doc? Am I going to get better? What's the treatment? Could the scanner be wrong?* We were told from our first day on the job that there was simply too much liability to let us make up our own

B. Aiken, *Small Doses of the Future*, Science and Fiction,
DOI 10.1007/978-3-319-04253-4_2, © Springer International Publishing Switzerland 2014

answers and that any violation of this policy would be considered a federal offense; this was, after all, a government clinic.

I looked at Ms. Johnson. "How do you know?"

"My friend Wanda is his assistant. She just texted me."

"She the one who turned him in?"

I felt the brief hesitation in her voice. "Nah, not Wanda. She'd never do something like that."

My attention was drawn back to Arnie as he snapped at a reporter, "And I'd do it again, God *damnit*. I'm sick of seeing my patients suffer just because I have to listen to some damn machine."

I knew exactly how he felt. We'd had that conversation over lunch at least a dozen times. The only reason that *I* had managed to stay out of trouble was because I didn't have the guts to do what Arnie did. I felt sorry for the poor bastard, but I admired him.

The crowd followed my beleaguered colleague out into the street where a black sedan was waiting. I hated myself for not trying to help, but what could I do?

We stood in the now sparsely populated lobby, staring at the scene on the other side of the picture window by the revolving doors. "Guess we'd better get to work," I said.

She gave a quick nod and we headed up to the 37th floor to begin our daily routine. By the time I got into my lab coat and made my way over to the exam room, she had already started the first medscan. Within minutes, a white plinth slid out from the mouth of the giant machine.

"Mornin', Doc," Mr. Winthorp greeted me, grabbing the back of his neck as he sat up from the exam table that had just emerged from the tube of the Medtron 3000.

Ms. Johnson looked up from the control monitor on the scanner. "No motion artifacts, Doctor. The report's coming up now."

"Thanks." I looked at my first patient of the day. "Good morning, Mr. Winthorp." I did not reach out to shake his hand. "I'm Doctor Jenkins."

He glanced up at the plaque on the wall displaying my diploma, barely legible behind a coat of fading yellow urethane. "Centerville class of 2012, huh?" He looked impressed. "Good school."

I hadn't looked at that piece of paper in a long time. "It *was*."

"So what are you going to do about my pain?"

I studied the report on the monitor. "The scanner has diagnosed you with a stomach ulcer and entered a prescription into the pharmacy system."

"Stomach ulcer? I got neck pain, Doc."

I pulled out my e-pad to consult the company manual and scrolled to the appropriate response grid. "I'm sorry, but the scanner says that your problem is a stomach ulcer. It doesn't mention anything about your neck."

"My stomach feels fine."

I scrolled further. Even though I knew most of the acceptable answers by now, it was best to be cautious, especially with a new assistant hanging on my every word. "Some illnesses have no discernable symptoms," I quoted.

Winthorp was too busy massaging his neck to notice that I was reading a script. "OK, maybe I do have an ulcer, but this damn neck pain is what brought me in here, not my stomach."

"Just the same, if you don't pick up your prescription, the insurance company will drop you from their plan."

Mr. Winthorp let out a huff through blowfish cheeks. He knew there was no point in arguing with a medscan. "OK, but can you just take a look at my neck? It's killing me."

The eyebrows on Ms. Johnson's fresh young face crested noticeably.

"I'm sorry, Mr. Winthorp," I recited dutifully, "but physical contact is strictly prohibited."

"Come on, Doc. I won't tell anyone."

My demeanor softened. "Now, Mr. Winthorp, you know I can't do that. I could lose my license."

He shook his head—with difficulty—and walked out the door.

I felt sorry for the poor sap. There was a time I'd have ignored the rules, taken a look at his neck. But that was before I watched a bunch of my colleagues go bankrupt from lawsuits for doing that sort of thing, or worse yet, get carted off in handcuffs like Arnie Hirsch.

But this was a new world. When I graduated from the prestigious Centerville Medical School 33 years ago, I couldn't have been more proud. Sir William Osler once said, "The transition from layman to physician is the most awesome transition in the universe." At least that's what we were told by our first clinical preceptor. And we believed him; thought we were special. After all, we'd gone from sniveling preppies to workaholics whose days were filled with making life or death decisions. That kind of thing changes a person. Changes you in ways you can't see, can't feel, can't notice until one day you wake up, look at on old picture of yourself and think, *Was I ever really that naïve?*

But it jades you, too. Rearranges your priorities. Makes it hard to maintain a normal sense of empathy, though most could; it's what made us good at our profession.

Or used to.

Ms. Johnson looked over my shoulder as I stood in the doorway watching Mr. Winthorp make his way out of the office. "Do you get many like that?" she asked.

"Nah. The scanner usually picks up the right thing; you know, whatever it is that's causing the symptoms."

"I can't believe that guy actually wanted you to *touch* him." She shuddered as she spit out the words.

I kept silent. The Board of Medicine was notorious for infiltrating practices with young trainees who were trying to weed out doctors who didn't follow the rules, and I didn't know my new assistant all that well yet.

She turned and looked at me. "I mean, I can understand how some of the older people might think that way; it's what they grew up with. But Winthorp's only 42. Why would *he* think a doctor could find something that a scanner couldn't?"

The poor guy was just looking for a little relief and we didn't give it to him; she had to see that. I wasn't going to fall for the bait. "Guess some people just long for nostalgia," I said. "Stories they hear from their parents, an old movie, some viral story running around the Web. There are lots of ways to hear about how things used to be. Some people still believe it was better back then."

"Are you one of them?"

I raised an eyebrow and used my slight height advantage to convey my answer without having to resort to an outright lie.

She seemed to accept that. "They don't know how good they really have it nowadays."

I nodded.

"It's just not analytical. Don't they know that people can make mistakes?"

"Spoken like a new graduate, Ms. Johnson."

Her lids narrowed. "You don't agree?"

"That people can make mistakes? Sure."

She shook me off. "That machines are the only way to examine a patient, that there's no need to ever touch a sick person as long as they can get into a scanner or surg unit by themselves."

I let out a deep breath. "Scanners are faster, more accurate and completely disregard emotion. No time wasted dealing with a person's feelings."

She looked relieved. "Exactly."

"But they can't empathize, can't connect to the psyche. There's a lot more to pain than nerve endings firing willy-nilly. The same pathology can cause different symptoms, different degrees of pain in different people."

Her eyes widened and I could feel my skin start to crawl.

I forced out a *hah* and said, "Gotcha."

Her demeanor eased, but her guard didn't drop.

"No, I've been down that road," I said. "You can't imagine the time it takes to deal with someone's feelings, much less the emotional stress that weighs on you. I'll take a scanner over that any day." I gave the Medtron 3000 a gentle pat on its cold, titanium side. "Thanks to these babies and the folks who came up with those answer trees, modern medicine has really evolved to a whole new level."

A relieved Ms. Johnson was back in amiable sidekick mode. "Makes the shift goes by pretty quickly, too."

I shot her a smile. I'd gotten pretty good at this game. I, too, had sunk to a whole new level; survival instinct is strong.

"Ms. Johnson?"

She turned.

"Bring in the next patient."

"Yes, Doctor."

Doctor. I sure as hell didn't feel like one anymore.

*

Another shift endured, I stepped out into the cool breeze of a late October evening and squinted up at the sunlight reflecting off the glass façade of the Unity Health Insurance building that dominated the downtown skyline, as I pulled my collar up and gripped it tightly against the wind. It would be a short walk home.

I'd been living in the city almost five years. The day Nan informed me that she couldn't tolerate having me around the house anymore, I decided to seek out an apartment within walking distance of the clinic. Oppressive crowds thronging in and out of mettube stations were not conducive to the mental well being of anyone, particularly those of my generation, and besides, I enjoyed taking in the…well, you couldn't really call it fresh air anymore, but I loved the atmosphere of the grimy city streets; preferred it to the sterility of modern buildings.

I made my way past *Hot Beanz*, my morning coffee spot; the aroma slowed my pace, but the thought of coming back out into the streets after warming up again kept me on my path. I turned the corner and approached the front door of my building, faced the camera embedded above the front door, and said, "Entry."

The oversized glass doors swished open and I hurried in out of the chill that my body would soon adjust to as the season progressed. I nodded at the animatronic receptionist in the lobby, which greeted me by name and summoned an elevator to the ground floor. As I entered, a perfectly pitched voice, the kind you hear on the 6 o'clock news, greeted me. "Going home, Dr. Jenkins?"

"Yes. Home."

"Very good." The door slid shut and I was escorted to the thirteenth floor, where I exited and made my way down the hall to the door where another *entry* command would grant me access to my little sanctuary.

I threw my coat over one of the checkered cloth-covered dining chairs, walked into the living room and looked out at the modest view of Centennial Park provided by the wall to wall windows that gave this place its charm.

"Play music," I commanded, just before flopping down into my favorite overstuffed black leather easy chair by the window; I pushed the little black button by my right hand and a foot rest popped up to the perfect height. "Petrushka."

As the music started to play, I closed my eyes and let it take me back to the day this album was recorded, a live performance in which my daughter had played the brief but famously recognizable trumpet solo the piece was known for, in her debut with the Chicago Symphony Orchestra. It had been one of the proudest days of my life; a life that was once filled with proud moments.

Medical school, marriage to my college sweetheart, three wonderful children, suburban bliss; all memories that now seemed more like someone else's life than one I had led myself. I should have seen it coming, should have noticed the signs, but I was blinded by the drive to succeed and failed to pay attention to the world evolving around me. The changes had been so gradual that they crept up on me like age, one wrinkle at a time. And then one day Nan asked me to leave. It wasn't really until that day that I realized just how *much* had changed; everything but me. The kids were all grown and scattered around the country; each a success in their chosen lines of work, but none a part of our daily lives anymore. Nan had managed to stay in sync with the pulse of the city; she had become a community activist, a prolific volunteer; she was doing things that mattered.

And I was but a shadow of what I'd been, increasingly disgruntled with a medical system that had long ago crumbled, a system that had lost its way from what it was meant to do—take care of people. I'd become so bitter that I was poisoning Nan's life, but never had a clue until the day she shattered my world.

It wasn't until that day that I realized Nan had been the one constant in my life that kept things real, that shielded me from the endless alterations reshaping the world around us; that *she* was the one who had been taking care of *me* all those years, not the reverse I had always taken as granted.

And on that day, I was lost.

*

After the divorce, it took me a few years to get a grip on life again; not joy, you couldn't really call it that, but I was beginning to discover things that fulfilled me, that gave me pleasure, that gave me a reason to live. I was starting to feel comfortable again until the day Arnie Hirsch got hauled away, then the questions of what I was doing with my life began to tear at me once more.

Stumbling in for my Saturday morning pick-me-up at Hot Beanz, an outing I looked forward to every week, a call of, "Jenks!" greeted me as I walked through the door. I hadn't been called that in a long time.

I looked up and smiled feebly. "Doug. How are you?"

Doug Barnes and I had gone to med school together and started a family practice soon after graduating. It was a thriving practice for a while, but the bureaucracy eventually caught up with us. Insurance companies only wanted to contract with doctors they could control, and we weren't willing to play the game. We thought we were better than that, but time wore us down. We were eventually forced to liquidate the practice and seek out clinic jobs like the rest of them. I hadn't seen him in years.

"Better than you from the looks of it," he said, waving me over to a table. "You look like hell."

I hadn't realized my desolation was that transparent.

We sat down, facing each other across a small round table. I smiled feebly. "Quite the coincidence bumping into each other here, huh?"

The edge of Doug's mouth curled up. "Nah, not really. It was Carma."

I peered at him over the rim of my glasses.

He waved his hands, and with a chuckle said, "Carma *Johnson*, your assistant."

"Really. Ms. Johnson?"

He gave me a nod. "She's one of *us*."

"Us?"

"Let me explain."

He went on to tell me that he'd been stuck in a clinic across town since we closed the practice, that he found it every bit as unrewarding as I found my job, and that the only reason he kept going in there every day was because he needed the money. Familiar story, but I still wasn't sure where Carma Johnson fit in.

Doug glanced around the room, then leaned in toward me. "Look, there's a group of us who get together every week. You know, people who feel the same way as you and me."

"And Ms. Johnson's one of them?"

He gave a single nod. "It's mostly physicians, but some nurses and techs have joined in too. We call it The Old Codger's Club, though it's been attracting a few of the more recent grads like Carma who thought they were getting into medicine for the same antiquated reasons you and I did."

"What the hell can you do besides bitch and moan to each other?"

"We run a clinic out of the back of a strip mall shop in the Libertyville area."

My eyes widened. The Feds didn't take to kindly to black market clinics.

"It's a nice blue collar neighborhood, not much crime, doesn't attract a lot of cops. We steer a few patients there, the ones we know we can trust. It's like the old days; we get to treat patients the way we were trained to instead of the way we're legislated to perform now."

"Jesus, Doug. What if you get caught?"

"Hell, it's worth the risk. Gives me a chance to shake the rust off, feel useful again. You should try it. We could use someone like you."

I knew exactly what he meant. You can only do so much pencil-pushing before you feel like you're starting to rot away. It was a tempting offer.

"How do you hide it?"

"Carefully. Don't talk about it to anyone you don't know, don't mention it at work even to those you trust. The walls have eyes."

"Tell me about it. Every time I get someone new in the office, I feel like I've got to spend all day looking over my shoulder. These kids coming out of school…they're brainwashing them young these days."

Doug laughed. "Carma got to you, didn't she?"

"Damn straight. I'd have sworn she was a mole for the Feds."

"Nah. Just feeling you out. Plays the part well, though, don't you think?

I had to agree. She'd figured *me* out without even a hint at what she was up to.

"So what do you say, Jenks? Our next meeting's tonight. Why don't you come check it out."

I rubbed at a stain on the table. I wanted to say yes, but I kept picturing Arnie Hirsch being dragged off in handcuffs.

"Well, at least think about it." Doug synced the info onto my PDA phone.

*

That's all I *did* do the rest of that day, think about it. Something he said had struck a chord. The idea of being part of a real clinic again made my blood flow in a way I hadn't felt in a long time.

I drove by the address Doug had given me. A quiet neighborhood strip mall. The storefront said *Fine Tailoring,* which I supposed was provided by a relative of someone in the Old Codgers Club. The information he had up-loaded to me included a password that would grant access to the clinic in the back of the shop.

I pulled up in front and sat there with the engine running as I stared blindly at the store. My car was relatively new, but no air conditioning would have been able to keep the sweat from soaking through my shirt. Office hours were from six till nine; I still had a few hours to make my decision.

I stopped by a Starbucks on the way home and grabbed a burger, cream soda and chips; carry-out bag. By the time I got back to my apartment, the food was luke-warm, but I preferred the confines of my home to a fast food joint. I wolfed it down, then jumped in the shower.

Most people sing in the shower; I think. In fact, it's where I do some of my best thinking. But even the hot steam swirling around me couldn't clear the fog inside my brain.

It would be so easy, I thought. *Drive to the strip mall, go to the clinic and get a chance to be a real doctor again.*

I pictured myself in handcuffs. *What am I, nuts?*

Hey, Doug's been doing it for God knows how long. How dangerous can it be?

Then a terrible thought occurred to me. *Maybe he's just setting me up.*

It's Doug, for Christ's sake.

Hey, I don't know what he's been up to for the last decade.

So what else are you going to do, rot away at Thirteen for the rest of your life? Show some stones, man.

I toweled off and glanced at the clock. Decision time.

At quarter after six, I left my apartment headed back to *Fine Tailoring*. My heart pounded faster with each turn and as I pulled into the lot the wheel slipped through my damp hands; only the car's proximity braking system saved me from plowing into a line of parked cars. I numbly listened to the electronic voice admonishing me for reckless driving until I had recovered enough to disengage the safety, then corrected course and crept along past the storefronts until I spotted an empty space directly in front of the tailor shop.

I hesitated, then tapped on the accelerator and turned out of the lot without looking back. A half hour later, I was home.

A bottle of wine kept me company that evening. I nursed it slowly, staring at the walls until finally deciding to go to bed whether sleep was in my immediate future or not. Dozing on and off, snippets of dreams flitted through my mind: med school, the old practice, nightmares of Carma Johnson walking in to my office with a team of uniformed agents. Doug had convinced me she was one of the good guys, but dreams don't always ride on facts and emotions don't erase that easily.

I was rattled out of my dreams a little after midnight by the shrill ring tone of an unprogrammed caller and stabbed out for the phone more in an effort to silence it than from any real curiosity about who was on the other end.

"Jenks? Jenks, that you? Why's your vid off?"

"I keep it that way when I'm in the buff," I rasped.

"Oh. Oh, yeah." I could see the stress lines around Doug's eyes as he looked down at his phone to check the time. "Jesus, I didn't realize how late it was. Sorry." He glanced back over his shoulder. "Listen, I don't know how much time I've got."

I squinted, trying to study his face through my blurry eyes.

"You were right."

"About what?"

"Carma. She turned us in. The cops raided our place tonight, just before closing. I had stepped out to take a break and when I got back there were half a dozen police cars out front. I've been trying to lay low, but you can only troll the streets for so long. It's just a matter of time…" I heard the sirens approaching his spot. "Jesus. Gotta go. Be careful, Jenks."

I reached for the remote control on my night stand and flipped on the monitor suspended from the far wall, then searched the web for local news. "Shit." There it was, plain as day. A bunch of doctors and nurses being hauled outside in handcuffs through the same door I'd been staring at only a few hours ago from the comfort of my car, the same door I'd almost walked through in a moment of rebellious false confidence.

"God, how could I have been so *stupid*? What was I thinking?"

I was too stunned to make out what they were saying before the picture faded to a live chase scene; Doug's car. I turned it off and tossed the remote back onto the table. I didn't want to watch the inevitable conclusion.

I flopped back and stared at the ceiling. My first glimmer of hope for a brighter, more productive existence in a very long time had been smeared all over the Net. All I had to look forward to now was District Clinic 13.

The phone rang. Doug's number again.

"Doug?"

"Dr. Jenkins?" A monotonal, unharried voice that was clearly not Doug's.

"Yes?"

A face came up on the screen, a generic clean-cut young male face adorned with a police cap. "This is Officer Harvey Cornell. Turn your vid on, sir."

I pulled a sheet around me and complied. Only my face would show on his phone, but it was still discomfiting to sit there with nothing on talking into a vid phone. "What's this all about, officer? Is Dr. Barnes OK?"

"He's fine, sir. Your number was the last one he called, just a few minutes ago, and we want to know why."

"Why don't you ask him?"

"We got his version, sir. We want yours."

I knew they'd review the transcript of Doug's phone call. *Don't' be stupid*, I reminded myself before answering. "He's my old partner. I ran in to him yesterday for the first time in years and gave him my number, so I guess it was at the top of his recent calls list. He sounded like he was in some kind of trouble. I guess in his rush to call someone he hit my number first."

"What do you know about the clinic, sir?"

"Uh, he told me about it yesterday, you know, when we were catching up on each other's lives." I fought against my instinct to wipe the sweat off my brow. The screen was small, maybe he wouldn't notice the gleam. I turned from the light.

"And you didn't turn him in?"

"I wanted to give him a chance to right it himself first. Warned him about one of his people, that she's a straight shooter. I guess he didn't take my advice, huh?"

"You'll need to come down to the station, sir. I'll be there in ten minutes to pick you up."

"But..." the line went dead.

Ten minutes.

Crap.

I threw on some jeans and a relatively clean shirt, brushed the stale wine breath off my teeth and paced in front of the door until the chime sounded, sending my heart crashing against the inside of my chest wall.

"Intercom on." The green light next to the door came on. "Hello?"

The animatronic receptionist from the lobby greeted me. "Good morning, Dr. Jenkins. There's an Officer Cornell here to see you. Shall I let him in?"

"Yes. Thank you."

"My pleasure, doctor."

I damped the sweat off my brow and rubbed the palms of my hands against my pants.

The chime sounded again. "Yes?"

"It's me, doctor. Officer Cornell."

"Front door, open," I commanded.

The door responded dutifully, and Officer Harvey Cornell entered with a vague scent of musk preceding him. A neatly pressed navy blue uniform accented his athletic physique, right down to the gleaming patent leather boots.

"Dr. Jenkins," he said, removing his hat and smoothing back the neatly cropped black hair held in place with a hint of gel. "Ready, sir?"

"Am I under arrest?"

"Not yet, sir."

"Then why can't we just talk here?"

He motioned to the door. "You'll want to come with me, sir."

Sometimes no answer is an answer you don't ignore.

The animatronic offered a cheery good-bye as we passed and made our way to the unmarked car waiting by the front entrance. A female officer sat perched in the driver's seat. Cornell opened the back door and I ducked in. He shut it behind me and I instinctively tried the handle, which of course did nothing.

On the way back to the station, he rode shotgun and didn't say another word to me. I could see the two of them conversing on the other side of the translucent barrier that separated us, but I don't know how to lip read. I only had the chatter of my own mind to keep me company.

As I sat there, every possible scenario flashed through my mind. Maybe they spotted me casing the clinic that afternoon, but that wouldn't be enough to arrest me on. They must have seen me pull up that evening, almost go in. But they can't arrest you for *almost*, can they? Hell, they didn't have anything they could pin on me. I'd been a damn boy scout at the clinic all these years; I hated myself for it, but I never gave them anything to hang me with. And what did they have now? My name in Doug's phone, a call, a drive by at the mall during clinic hours? Nothing. They had nothing. Still, they could make my life miserable if they wanted. I'd been a damn poster boy for the District Clinic System, ignored what I knew was right to spite the health of my psyche, and they were going to screw me anyway. Great.

The flashes of panic were knocked from my thoughts by the sound of the car coming to a stop. We were parked outside the station. Cornell opened the door and escorted me into the building, where we wound our way through a maze of busy cubicles and into a sealed interrogation room. There was no mirrored glass, but there was no doubt we were being recorded.

He sat across a polished steel desk, facing me, but staring intently at a computer screen to his right. His face remained expressionless as he read silently and periodically tapped on the screen.

I cleared my throat, quite unintentionally, and was speared by a 'don't do that again' look from across the table. A few minutes later, Officer Cornell sat back against his chair.

"Doesn't look too good for you, doctor."

"What doesn't look good? What are you accusing me of, being friends with Dr. Barnes?"

"You should be more careful who you associate with."

"Since when did that become a crime?"

He stared me further back into my seat, then stepped out of the room. I squinted in all directions trying to locate the camera. *Christ, they can't lock me up just for thinking about going to that damn clinic, can they?* I pulled a tissue out of my pocket and damped off my face. *Stay calm*, I coaxed, but my body wasn't listening. I tucked the fraying wet tissue into my pants pocket as the door popped open and Officer Cornell re-entered.

He sat down and tapped on the screen, looked at me for an excruciatingly long 3 or 4 seconds, then focused his attention back on the monitor.

I scooted around on the cold steel seat of my chair in a futile effort to get comfortable.

Cornell looked up again. "Look, doctor. Let me be blunt."

Finally. I'd have rather been arrested than have to sit in that seat any longer, staring at the machine that called himself Officer Cornell.

"We've got video surveillance that shows you hanging out in front of Barnes' clinic this afternoon, and then driving by again tonight, just before we got there."

I could feel the heat rising up from under my shirt and thanked my lucky stars he didn't have me hooked up to an autonomic monitor to graph my anxiety. Not that he needed one.

"He was my friend. I was just curious."

"Don't insult me, doctor."

I opened my mouth, but nothing came out.

"Look, we may not have anything damning on you, but with the video, the phone call, your connection to Doctor Barnes, well let's just say it's pretty clear what your intentions were. You were more than a little tempted to join his party, weren't you?"

Before I could answer that every-chamber-loaded question, he stopped me. "You were lucky as hell tonight, but don't count on luck to strike twice. That space you have been flying under the radar in has just gotten considerably smaller."

The tension permeating every fiber of my being had begun to ease. They were going to have to let me go. "So I'm your new assignment?"

"Even if I had the time to stay on your ass, which I don't, I don't believe in entrapment. But I'm not the only one with this information. Consider tonight a friendly warning."

This kind of friendship I could do without. I felt a chill as the sweat began to cool against my skin.

He stood. "You can see yourself out. I've got to get started on those damn reports. That's the penalty for working with the Federal Health Care Task Force; paperwork's a killer." He pointed the way out. "We can have someone drive you home if you'd like."

"I'll cab it, thanks."

"Thought you might." He started to walk toward a cubicle to the right of the interrogation room, then hesitated and turned. "Be smart, doctor."

I couldn't get out of there fast enough.

*

Monday morning came, as it inevitably did. I made my way up to the thirty-seventh floor where Ms. Johnson was waiting for me.

"So you didn't take the bait," she said wryly.

"No. I did not. You know I don't go for that sort of thing."

"I do now, but I could have sworn you'd go for it and I'm usually pretty good at reading people."

My task had become doubly burdensome. I felt like I was working under even more of a microscope than I had before. But I endured. And I thought about Doug. Constantly.

A couple of weeks later, Mr. Winthorp came by again on a Friday afternoon, still in pain and still begging me to examine his neck. There was nowhere else for him to go; I was his assigned provider. Once again, I turned him down and it ate me up inside.

It was getting harder to look at myself in the mirror, harder to accept what I'd become after seeing that there was another way for those willing to do what needed to be done. Sure, they'd caught Doug, but there were dozens of clinics that managed to stay under the radar, if you believed the blogs. I never had. I desperately wanted to now.

Still, it was tough to ignore Officer Cornell's warning.

That night, the pros and cons played in my head a hundred times over as I lay in bed praying to be mercifully overtaken by sleep. For once in my life, I had a decision to make for which I wished there *was* an answer tree to guide me.

At 2 AM, I awakened with a bolt. "Doc Tramer's place...*of course.*" The image was plain as day now, the obituary from last Saturday; my old family doctor, the one who used to see me at the office in his house had passed away at the ripe old age of 97. A paragraph of accolades and a statement expressing how sad it was that he had no survivors; his house would be going up for sale.

I pulled up the number of an old realtor friend of mine first thing in the morning, then jotted it down and left it on the table while I made some coffee. As I munched on a bagel and sipped my java, my gaze kept straying from the news on the monitor back to that little scrap of paper.

But they're watching you.

Bullshit. You really think you're that important? They don't have time to bother with you. It was just scare tactics.

You willing to take that chance?

I tilted my cup to get one last rush of caffeine, then started to rise from the table. "Ah, *hell.*" I spun back and grabbed the note.

*

The realtor was already waiting in the driveway when I pulled up to the old Tramer place. Doc had been retired for a couple of decades, but his home office was still intact; a veritable shrine to the medical era I grew up in. It looked like he'd taken a lot of pride keeping it that way until the past few years when he'd undoubtedly had to occupy his time just trying to survive.

It was perfect. The office had been out of commission long before the District Clinic system was a glimmer in the eye of the jackasses who created it; the Feds wouldn't even know this place existed.

A scent of mold hung in the air and the house looked like hell: faded paper peeling off the walls, archaic appliances, incandescent light fixtures; a realtor's nightmare. But mostly cosmetic stuff I could deal with myself. I made an offer on the spot. She couldn't get the contract to me fast enough.

*

Weekends had always been my cherished time, the outdoors my playground. Whether it was people-watching in town, or escaping to the little park land that remained within commuting distance, I'd spend my days trying to commune with the things that made life worth living.

But now, I had the perfect retreat. The quaint house was nestled next to a neighborhood park, with a beautiful view of the foliage from the second floor master bedroom window. I began spending my weekends there and the renovations went quickly. Within a month, I was ready for my first visitors.

Boredom and security were about to be replaced by fulfillment and paranoia.

I had kept the décor very retro. Faux-oak paneling warmed the walls in the foyer; the leather sofas were real. I admired my handiwork as I prepared for my first Sunday afternoon clinic. Taking a page from Doug's failed attempt, I was determined to fly solo on this.

Easing back into a well worn sofa cushion and relishing the faint moldy scent of the period throw rug scavenged at a flea market, I folded back the sports pages of the January issue of the New York Times, the last newspaper still printed in hard copy. The monthly edition didn't even try to keep up with the kind of breaking news coverage you could get on the Net, but the in-depth human interest stories were compelling, and there was no substitute for the satisfying feel of brittle pages of newsprint crinkling through your fingers.

The nostalgia of a simpler time, a more *humane* time, soothed my soul.

As I sat enjoying the moment, a mellifluous chime reminiscent of the period redirected my attention to the double front doors, where an adjacent monitor lit up with the familiar face of Mr. Winthorp.

I smiled and buzzed him in.

3
Freudian Slipstream

OK. This one's just for fun...mostly. But it does take a look at a very serious problem we may face as our technology sends us further into space. I won't say what, because that would spoil all the fun.

So relax, and if you can, read this on a beach somewhere with an ice-cold drink in one hand as you run your toes back and forth through the fine white sand.

And if you're not fortunate enough to be lounging on the beach, imagine yourself sitting on a stool at an open air bar by the ocean somewhere in the Caribbean. You're sipping on a beer, a Pina Colada, maybe a seltzer with a twist of lime, if that's your thing. Palm trees are swaying, the surf gently brushing against the shore, and every now and then a sea gull calls out as it swoops by looking for a handout, a salty pretzel or peanut, maybe. No one in sight except you and Sam, the bartender...

Jackson Carr looked at his half empty mug, then gazed back down the length of the desolate open air bar on the beach where he'd arrived just an hour earlier. Waves of heat radiated off the sand, and through the blur a vaguely familiar image began to take form.

Shadowy fragments wavered above the sand and surf, gradually coalescing into the rhythmic figure of a jet-black stallion loping effortlessly along the shore. But it was the rider that caught Carr's attention, her long blonde hair rippling through the ocean breeze as she moved in perfect synchronization with the animal.

She rode up to the beachside bar, gliding off the horse's back as it eased to a stop, then grabbed a steel bucket from behind the bar and placed it on the counter.

"The usual, Sam," she said to the bartender, a hint of southern twang sneaking through.

She looked over at Carr, the sole patron of the establishment until her arrival, and gave him a smile. Her white lace camisole billowed in the salty air just above the hip line of her faded jeans.

B. Aiken, *Small Doses of the Future*, Science and Fiction,
DOI 10.1007/978-3-319-04253-4_3, © Springer International Publishing Switzerland 2014

Carr sat balanced on a barstool under the shade of the thatched roof over-hang, listening to the rustling palm fronds and nursing a beer as he studied her face. The finely sculpted features looked so familiar, but he couldn't quite place her. Probably a model he'd seen on the Net or some billboard back in town.

Sam poured a glass of ale from the tap, then filled the bucket to a froth and set it down in front of the horse. The stallion snorted in thanks and lowered his head into the pail.

"You're welcome," Sam said.

The woman let out a laugh and gave the horse a pat on the neck, then picked up her glass and drank in one long slow draw until the last drop of ale was gone.

"It's a hot one today, Sam." She pressed the glass up against her cheek, revel-ing in the coolness, then ran it down along her neck.

Sam continued cleaning up behind the bar without giving her a glance. "That it is."

She placed her glass back on the counter. "For a bartender, you sure don't talk much."

"Only when need be."

"Some people just don't know how to have any fun." She winked at Carr, fixing him with emerald eyes, then swung up into the saddle and tossed back her hair. With a flick of the reins, she cantered past him, continuing her jour-ney down the beach as he burned the image into his mind.

Once she had faded to a dot on the horizon, Carr looked back at Sam and let out a slow breath. "Man, I've been living in the wrong place for the last twenty-nine years." He ran a thumb up and down the foggy layer of con-densation on the outside of his mug, shook his head slowly as he studied the pattern. "All that time slaving away in an immunology lab…" He looked up and scanned the horizon, took in a deep breath of the thick salty air. "I haven't been to the beach in ages. Hell, I've barely been outdoors."

"Nobody has, friend. Too damn much radiation." Sam picked up the emp-ty glass at the other end of the bar and wiped the countertop where it had been sitting, then flipped the tattered rag over his shoulder and grabbed a straw broom that had been leaning in the corner.

Carr gazed in the direction the woman had ridden off. "Does she come around often?"

"Every day around this time."

"Who is she? She looks so familiar."

"No one you know."

Carr let out a half grunt, half laugh. "Have to change *that*."

Sam stopped sweeping the sandy floor and studied Carr's face. "You're not ready yet."

"Are you kidding? I'm way past ready."

"Stay focused, Jackson. You've got work to do." Sam turned away, tossing a bowl of stale pretzels onto the adjacent pristine beach.

A seagull called to its friends as it spotted the prize and soon a chorus of whistles overtook the afternoon peace.

*

Scattered fragments of memories rattled around in Carr's dreams, but nothing seemed to come into focus. Bits and pieces of a briefing about a world called Carpathia, images of colonists arriving on a lush planet, but not much more. He couldn't shake the feeling that those colonists needed his help, though he couldn't imagine why. He'd never even been off Earth. It didn't make any sense. He tried not to think about it.

*

The sun was high overhead when the lady on the horse appeared on the horizon. Carr was already at the bar anticipating her arrival. She slid gracefully off the back of the stallion and made her way over to the bar.

Sam filled the bucket from a blue acrylic keg behind the bar, and topped off a mug of the same.

Once again, she finished her beer in one easy motion, then smiled at Carr on her way out. He was sure the smile was more personal this time.

The two men watched her ride off down the beach.

"You ready to work yet?" Sam asked.

"I want her, Sam."

"You're not ready."

Sam handed Carr a mug of beer from the acrylic keg.

Carr took a sip without diverting his gaze from her path, then spit it to the ground and slammed his mug on the bar.

"What is this piss?" he snapped.

"It's what you need."

Carr dumped it in the sand. "Gimme a Beck's."

Sam complied and Carr took a long draw from the frosted mug, eased back in his seat and peered out across the cool blue waters.

"I've been having the weirdest dreams, Sam."

"Weird how?"

"I don't know," Carr sighed. "They seem so real, but they don't mean anything."

Sam stopped mopping the counter. "What are they about?"

"You'll think I'm nuts."

"Try me."

Carr hesitated, then looked Sam in the eye. "Ever hear of a planet called Carpathia?"

"Sure. It's been all over the news."

Carr was confused. "What news?" He motioned around the rustic bar. "I haven't seen a vid screen since I've been here."

Sam's gaze didn't wander from Carr's face. "It's time to remember, Jackson."

Carr shook his head. He didn't want to remember anything that might take him away from this paradise. He looked out to the sea and tried to clear his mind. Waves crashed to the shore renewing the life that crawled beneath the wet sand, and the afternoon sun burned through the mist.

*

Try as he might, Carr could not stave off the images that danced in his head, trying to coalesce into meaningful ideas. Strange thoughts haunted his dreams; people on a distant planet crying out to him for help, but why? He was no hero, just a man who spent his days slaving away in a lab.

*

The next day, Carr sat down in his usual spot, looking down the beach in the direction from which he knew she would come. But this time, tranquility refused to wash over him.

"You know what they mean, don't you, Sam?"

"They?"

"The dreams. Carpathia."

Sam nodded without looking up from the broom he was pushing along the floor behind the bar. "So do you."

"If I knew, why the hell would I be asking?"

"Think." Sam paused from his chores and leaned against the wooden handle. "The news reports…global warming…the briefings. It's time to remember."

"Look," Carr snapped, "I told you, I…" He stopped abruptly and his eyes glossed over as some of the fragments of memories started to gel.

Sam smiled as Carr closed his eyes and nodded slowly.

"Carpathia," Carr muttered, "The new Earth. So they finally found a way to save our skins, huh?"

"Good," Sam said. "You remember."

"Enough to know why I forgot. What a stupid idea. Escape the Sun's radiation by hopping over to some planet twenty-two light years away."

"It's the only way."

"Two years in a cramped little spaceship? No thanks. I'll stick to sunscreen." Carr took a big swig of his beer. "Besides, what the hell do they need *me* for?" He looked longingly down the beach. "I'm not leaving here, Sam."

"You already have," Sam said, turning away.

Carr looked at him curiously. "Always riddles with you, my friend. Always riddles."

The muffled sound of hooves pounding into the hard-packed sand floated in on the wind. Carr looked up as the lady on the horse approached, and Sam filled a glass from the keg with that God-awful brew.

She downed her beer quickly, and then left with an inviting smile. As she rode away, Sam again slipped him a glass of the dark, frothy ale and he drank.

"God!" Carr yelled. "What is it with you? Is this some kind of sick joke?"

"No joke," Sam said. "It's what you need."

Carr tossed the liquid onto the sand and thrust the mug at Sam. "I'll tell you what I need. I need something to chase that filth from my mouth."

Sam complied. "You have work to do," he said as he slid the mug across the bar.

Carr reached for the beer. "Lighten up, would you? What's the rush?"

"Time is not what it seems, my friend."

A tropical breeze whistled through the fronds of the coconut palms behind the bar, cleansing the air.

*

Carr began to remember the early reports from Carpathia more clearly. The climate near the equator was tropical and the indigenous species of birds and fish were surprisingly similar to those found on Earth. The colonists had established a small village on a mountainous landmass. The pictures they sent back were broadcast all over the world. It looked like paradise.

*

For the past 2 weeks, Carr had spent his days at Sam's bar on Sunset Quay watching seagulls float on the breeze and sipping cold beer while awaiting the arrival of the lady on the horse. It never struck him as odd that he and Sam were always the only ones there.

Each day, she repeated the same routine, and each time she smiled, he could feel himself being drawn closer.

"Do you know where she lives, Sam?"

"In a place you shouldn't be."

Carr scowled. "I'm getting tired of your attitude, man. What's it to you anyway? You got a thing for her?"

Sam bellowed with laughter.

"What? What's so funny?"

"You haven't really seen her yet, have you?"

"Enough to know that I'd like to see more."

"Careful what you wish for," Sam started.

"Yeah, yeah. I know. It might come true," Carr sniped. "If only."

Sam fixed his gaze on Carr. "You've got work to do, Jackson."

Carr sat quietly for a moment, eyes fixed on the ebbing tide, then looked back over the counter at Sam. "All right. Let's get to it."

Sam nodded and they walked around toward the back of the building.

*

Over the past few days, Carr's memories had started to come into focus. One particular image, the last transmission he'd seen from Carpathia, was vivid.

Video was sent of the Carpathian flyers, indigenous creatures about the size of fruit bats with cat-like fur and big green eyes. They were alluring little things with an affinity for human contact, but their touch was far from comforting. Within seconds, skin began to melt away from anywhere there had been direct contact with the soft fur, and the ensuing death was a welcome relief from the horrific searing pain that spread like napalm oozing across flesh. It was an image Carr would have preferred not to remember.

The colony's physician had localized the cause, a potent toxin secreted by microorganisms in the fur. In a symbiotic relationship, these organisms thrived on nutrients in the hair follicles, and they in turn kept away predators that had painfully learned to avoid contact with the flyers.

The colonists raised a force field around the inner colony, one as large as their power cells could support. They were safe for now, but sooner or later the power would fail.

The message sent back to Earth included a detailed description of the offending organism. Thanks to quantum slipstream technology, the message arrived in minutes despite the great distance, but it would take nearly 2 years to send reinforcements; the technology was far more effective for sending optical data bursts than it was for starship travel.

Time was of the essence. The power cells supporting the force field would only hold out for a couple of years, at best. A solution would have to be found while on the way.

*

Sam led Carr around behind the Tiki bar where he'd been spending his days. It was a picturesque wooden structure with thick palm tree beams and a thatched roof, set directly on the beach and abutting the mountains behind. He tugged on a cracked wooden door, swollen from the humid air. It gave way on the third pull and creaked open.

Carr followed Sam in.

He should have been amazed, but he was not. The room was washed in bright fluorescent light and somehow seemed larger than the entire building had appeared from outside. Gleaming white linoleum covered the floors and the countertops were formed from black stone, polished to perfection. Glass-front cabinets were lined with test tubes and beakers, and all the familiar reagents Carr had been working with for the better part of his career. It was an exact duplicate of his research facility at the university.

*

Carr still wasn't sure where he was or why he had chosen this particular beach, but he remembered preparing for the trip and being briefed about the rigors of interstellar travel.

Shortly after the first mission had left for Carpathia, concerns surfaced that supplies would be insufficient if any unforeseen problem slowed the voyage. It was a chance the colonists were willing to take for the opportunity to go to the new world, but an unacceptable one for most of those who would follow.

The only viable solution would be to keep the passengers in suspended animation to conserve resources, but prior attempts had failed miserably; any period of stasis longer than a few days resulted in atrophy of the body and mind, effectively turning people into little more than organ harvesting material.

Efforts were redoubled to find a solution. Muscles could be exercised with electrical stimulation, but stimulating the brain proved to be more of a challenge. To combat the problem, engineers developed stasis chambers controlled by artificial intelligence computers. Each computer, after extensive conversations with its future suspended animee, could develop a dream-state life in which to engage the subject during the journey, thereby keeping the traveler's mind active throughout the period of stasis.

The technique proved successful and they were able to transport a large number of individuals to Carpathia to assist the colonists.

Jackson Carr was one of those people.

*

"How did you do it, Sam?" Carr methodically perused the sterile looking room. "If this place wasn't so damn clean, I'd swear I was back at my own lab."

"Of course," Sam said. He activated a viewscreen that dropped down from the ceiling directly over the main workbench in the center of the room. A series of three-dimensional models of organic molecules appeared across the screen. "Here's what I've got so far, but I can't get any further without your help."

Carr looked at the screen, carefully studying the molecular structure of a designer T-cell that was being fabricated to attack a very specific pathogen, one which was depicted in the far right lower corner.

"Where did you come up with this? It's brilliant."

"I got it from you, Jackson."

"Thus, the brilliance."

"Credit where credit is due. That's why you're here. Your paper in The Annals on targeted T-cell fabrication for silicon-based pathogens. The one you based on the early uploads from the Carpathian Immunologists. I've taken the liberty of mixing up the formula based on your work, and adjusted the program so that you'd taste a sweetness if the T-cells bond properly."

"Guess you got something wrong, huh?"

"Apparently so."

Carr moved toward the model of the pathogen to take a closer look. His jaw dropped open. "How could I have missed this?" He tapped on the screen and enlarged the image.

"Because it wasn't there, Jackson. The structure has changed from the one you saw when you wrote your paper. This one's based on new data we received from Carpathia since you did your original work."

"Imagine that," Carr said. Carbon atoms infused into the silicon matrix to help it bond to human tissue, make it a more effective killing machine. The buggers must have adapted since the colonists ran the first scans." He remembered the graphic images of the first colonist killed by one of the flyers. "As if the original bugger wasn't nasty enough."

Sam stood beside him, looking at the complex structure of the alien pathogen. "So what do we do?"

"Change the configuration of the bonding bridge on the T-cells. It'll take some time."

*

For the next 2 weeks, they started off each day in the lab. Carr knew what to do, but it was tedious work. By late morning, he would need a break. Besides, Jackson couldn't concentrate when he knew *she* was coming.

The lady on the horse appeared each day at her usual time. And each time she arrived, Carr was in his seat at the end of the bar taking it all in.

This went on for weeks, and each day she smiled a little more seductively without ever saying a word to him. Her routine never changed, not until the day Carr had finally completed his work.

"That it?" Sam asked, as he looked at the sealed Erlenmeyer flask Carr had just set on the counter.

"Yup. It just needs to set at room temperature for at least two hours. Once the reaction's done, we mix it in with the formula you whipped up and it'll alter the T-cell binding sites to make them compatible with the pathogen. It will be almost instantaneous."

Sam nodded. "Good work, Jackson."

Carr glanced up at the clock on the wall. "It's almost time." He threw off his white coat and headed out of the lab to settle in to his usual spot at the bar.

*

He sat with elbows propped on the sandy counter, exhausted, but felt renewed as he watched the object of his fascination quench her thirst once again. She gave him that smile, then lifted the reins and started to ride off. But this time she stopped as she approached the spot where Carr was sitting and extended her hand. He took it and mounted the horse with ease, settling in behind her.

Sam's warnings faded with the wind.

*

Sam hoped he had chosen the right path. Watching Carr ride away, he had his doubts. He searched his memory for any possible flaws. There had to be a way to focus Carr's thoughts without pulling him out of this world.

Each passenger who had been chosen to make the voyage to Carpathia was assigned an AI. Jackson Carr was Sam's first assignment since joining the relocation center. As per protocol, Sam had probed Carr's personality, life experiences and knowledge base down to his innermost thoughts. This process enabled Sam to garner enough information to generate a program designed to keep Carr in good health while allowing him to work on the assigned problem. In Carr's case, the assignment was to develop an antibody to protect the colonists against the deadly organism they'd come into contact with on the new world.

In preparation for the journey, Sam had questioned Carr for two grueling days. As fatigue set in, Carr started to let down his guard. He related an intimate story that had captured his fantasy in adolescence, one that had held his imagination hostage for a lifetime of dreams.

Once Sam had processed Carr's emotional reaction to this story, he knew the image would serve his needs.

*

Carr held on to the lady on the horse as they rode into the wind. Within minutes they arrived at a mountain villa overlooking the ocean. She took Carr's hand and led him up a stone path to the villa. It had been mid-day when they started up the path, but the sun was setting by the time they reached the top a few moments later.

Without a word, she led him up a spiral staircase and into a large room decorated in Renaissance era furnishings. Lavish tapestries adorned the walls and two divans upholstered in red velvet were positioned on either side of the large picture window overlooking the beach. The matching red velvet curtains were fastened open with golden ties on either side, allowing sheer white drapes to billow in and out of the window on the ocean breeze.

Carr turned to see her, clad in a white lace robe, reclining in the elegant four poster bed and summoning him with the index finger of her outstretched hand.

The evening far surpassed the memory of his adolescent dreams and he drifted into the most peaceful sleep he had ever known.

*

Sunlight warmed the heavy lids of his eyes and Carr awakened to the sound of gentle laughter. He turned to see her watching over him and met her serene smile.

She pulled the hair back off her face, and Carr went numb as that perfect skin began to stretch, distorting itself into a sinuous mass of gelatinous fibers that radiated out in all directions around her eyes. Her laughter had morphed into a macabre bellowing that could not possibly be coming from the woman he had made love to only a short while ago.

The sheets fell away from her body, revealing an amorphous bloated figure barely supported by its receding limbs. Carr leapt off the bed in horror, more from the thought of what he had done the night before than from the actual sight of what was in front of him. He pulled a sheet around himself and backed toward the window, unable to divert his gaze as she continued to change.

Now little more than a fusiform blob covered in golden fur, it floated up from the bed and closed the distance between them. Carr stumbled over the sheet and fell just as the creature burst open. He raised his arm to protect his face, but not before seeing hundreds of Carpathian flyers emerge from the shards of distorted flesh and glide out the window toward the bar on the beach.

There was an intense burning in his belly, and Carr doubled over in spasm. He was sure he had only minutes to live before the alien pathogen would dissolve him from the inside out. Forcing his body upright, he pulled on his pants, then darted out of the villa and down the stone walkway to the beach. He jumped from the edge of the stone wall at the base of the stairs onto the waiting stallion and galloped off towards Sam's place faster than the wind that carried the alien flyers toward their target.

Sam was in front of the bar watching the approaching gray cloud of Carpathian flyers, his right hand shielding his eyes from the rising sun overhead, when Carr arrived.

"You weren't ready," Sam said.

Carr stared in disbelief for a brief second. "No time for lectures, old man," he snapped. "Let's finish it."

The flyers' shadow began to encroach as they drifted between the angle of the sunlight and Sam's bar. Carr could hear the shrill call of their voices as they approached, indistinguishable from the laughter of the beast from which they'd emerged.

Sam glanced up, then calmly turned and walked to the lab behind the Tiki bar. Carr followed on his heels.

As the door sealed shut behind them, they could hear the flyers scratching against the wood.

"Time is short," Sam said.

Carr nodded. He knew the flyers would find a way in soon. He reached for the Erlenmeyer flask and poured its contents into an acrylic blue barrel sitting on the counter. A fizzing noise from within could barely be heard over the ever increasing squeals of the invaders trying to breach the lab.

Sam smiled. "That should do it."

He raised the spout attached to the barrel by white flexible tubing and shot a squirt directly into Carr's mouth.

Bracing for the rancid flavor of the frothy brew Sam had tried to serve him many times before, Carr was relieved by what he tasted. "It's sweet as sugar."

Sam smiled.

The burning in Carr's belly was gone and the sound of the flyers subsided within seconds.

"What's happening?" Carr asked.

"You're done, Jackson. That's all there is." As he spoke, Sam's voice drifted away, and with it the bar and the beach and all that surrounded them melted into nothingness.

*

It was worse than the worst hangover he had ever had. In spite of the best efforts of his AI, nearly 2 years of sleep in a state of suspended animation had taken its toll.

Carr peered through his barely open lids trying to get his bearings.

"Sam?" he called out. "Where are you, Sam?"

"My name is Austin 14, Dr. Carr," the voice said. One of the ship's robots was standing over the stasis chamber Carr had been sleeping in. "We have arrived at Carpathia. The captain is awaiting the results of your research."

Carr's mind slowly began to clear. He wasn't sure where the dream ended and reality began, but he remembered the mission and he remembered the formula for the antibody.

"Of course," he said.

He tried to sit up, but almost passed out from the dizziness. Austin 14 eased him back down and pressed a button on the console adjacent to the stasis chamber.

"I am administering a vasopressor to counteract the orthostatic hypotension induced by your prolonged recumbency," the robot announced.

"Huh?"

"I gave you something to stop the dizziness. You can sit up now."

"Oh." Carr said, and he slowly sat forward testing the results of the robot's treatment.

This time, he felt fine and his head was beginning to clear.

"Thanks, doc," he said.

"I am not a medical robot," Austin 14 said, "merely a ship's steward."

"Thanks just the same." A smile creased his face as he slid off the plinth.

"You are welcome."

"Right," Carr said. "Better get me to the captain. I've got the formula for the vaccine to protect the colonists."

"The captain will be pleased."

A wide grin stretched across Carr's face. "Just wait till he tastes it."

4

Done That, Never Been There

First printed in *Analog Science Fiction and Fact* in September, 2012

It's tough to solve a case when you're 240,000 miles from the scene of the crime. Nuff said.

I'd cleared my schedule for the rest of the afternoon; didn't want the distraction of patients waiting for me while my interview with that annoying little dweeb of a reporter dragged on past the scheduled time, as it undoubtedly would. The office door was ajar as I signed off on some of the endless charts backlogged on my system while I waited for him to show. It was a welcome relief when my secretary, Doris, poked her head in.

"Mr. Green is here, Dr. Bennett." She gave me a knowing wink with those familiar green eyes that used to drive me wild, but now only kindled the comfort of familiarity. Still, she carried her fifty-six years much better than I carried mine, and her youthful vigor helped me deceive myself as long as there were no mirrors around.

I tapped on the pad in front of me, and the image suspended over my desk disappeared back into whatever genie-box the IT guys had it programmed to pop out of. "Show him in."

She cleared her throat and motioned toward her neck as if she was straightening an imaginary tie.

"Ah, right." I buttoned my top shirt button, snuggled-up the brown and blue paisley tie that had been hanging loosely around my neck, and pulled down on the lapels of my white lab coat.

She gave me a nod of approval, then disappeared behind the door. A few seconds later a tall stick of a man, forty-something with short-cropped wavy red hair, came bounding into my office, then paused just inside the door. "Sorry I'm late." The apology came from a face full of freckles and the kind of milky white skin you can never expose to the sun for very long. He waited an

First printed in *Analog Science Fiction and Fact* in September, 2012

B. Aiken, *Small Doses of the Future*, Science and Fiction,
DOI 10.1007/978-3-319-04253-4_4, © Springer International Publishing Switzerland 2014

awkward moment for acceptance that never came, then hurried over to my desk and extended his right hand. "Tarin Green."

I curled the edges of my lips as far down my chin as I could force them; didn't reach out for his hand just yet. "Who the hell are *you*?"

A brighter red filled the spaces between his ruddy freckles. "I…uh…we… have a meeting, an interview. Don't we?"

I've had half a dozen interviews with prominent reporters over my career and it's not easy to make them sweat, but I seem to have succeeded rather quickly this time. The glow of his cheeks brightened as I enjoyed the moment.

He stammered, "My, uh, secretary said you'd agreed to…"

My laughter broke the mold of the scowl I had been trying so hard to maintain. I stood and reached out to shake his hand. "That, I did. And we do. Though for the life of me, I can't imagine why."

He stared blankly at my hand for a long instant, then took it. His grip was cold and clammy, but his handshake was vigorous.

"You're a hero, doctor. A genuine hero."

"Uh, thanks."

I pried my hand loose and relaxed into the high-back swivel chair that made doing all my e-charting somewhat less intolerable. I couldn't help but notice Green rubbing his hands along the sides of his pants legs, right index finger oddly extended, as he took a seat in one of the plaid cloth-covered chairs across the desk from me, then patted at the right chest pocket of his beige tweed jacket. Awfully jittery for a man that had no doubt been exposed to far worse than the meager prank I'd just pulled. "You OK?"

His head jolted up. "Hmm? Oh…yeah, fine. It's just, well, I've been looking forward to meeting you so much. To actually interview someone involved with the Moon Base program…"

I looked him straight in the eye—the right one, if you must know; always had a preference for the right one for some reason I refuse to analyze. "You do realize, Mr. Green, that I have never actually *been* on the moon."

"Well, yes. But, well, it's really just a matter of semantics, isn't it? I mean, it was you who performed the first major surgery there."

"Sitting in *there*," I pointed to a walnut door to my right, "strapped into the VR chair that controls the surgical robot bolted into the floor of the operating theater on MBA."

"Just the same, you *were* the one who had the guts to tinker around in somebody's brain from 240,000 miles away."

What I had done was no great secret. It had garnered far more press than any robotic surgery procedure should have warranted. It wasn't really any different than the remote surgery we do for cruise ship passengers, Arctic explorers, and those who choose to live their lives in the solitude of remote locations all over Earth. But the media had made me out to be some kind of hero: the man who convinced everyone that they could safely venture to the moon. As if the lack of a surgeon was the only danger. But after those stories were plastered all over the Net, NASA had no trouble getting workers to sign up to go to the moon and start construction on Moon Base Alpha.

When I'd performed that first operation, there were twelve hearty souls paving the way, living out of a dome no bigger than my apartment. Since then, MBA had grown to one hundred and twenty-three residents—a diverse group that was beginning to resemble a small town, though still housed in one large central complex.

Green swayed subtly in his chair, a motion no doubt caused by the habitual twitch of one leg crossed over the other, the sight of which I was mercifully spared by the desk between us. It seemed an odd habit for such a seasoned reporter.

"My readers want to hear about that first operation. What was it like?"

As I was about to answer his question, I couldn't help but notice something was missing. "Aren't you going to record any of this?"

"Oh…uh…" He reached into his pocket and pulled out a flash recorder. "Right. Thanks." A quick laugh forced its way through his rigid lips. "You'd think I'd never done this before." He tapped the recorder on, then gave me a nod. "So what was it like, that first operation you performed on a patient on the moon?"

I took a deep breath. I hadn't asked to be the center of attention. I could do just fine without the fame, thank you. But as much as I wanted to kick this freckle-faced nerd out of my office, I knew that the only thing worse than the notoriety he was offering me was the venomous infamy I'd receive from his substantial group of fans if I treated him like the pest he was.

"Pretty much like what I do any other day of the week," I said. "I'm a neurosurgeon. Tinkering around in people's brains is what I do. We use robotic surgical assistants in the operating room all the time; that gadget in the other room isn't too much different. The only real difference is that if something goes wrong, you can't jump out of the VR chair and finish the job the old fashioned way."

"No room for error, huh?"

"Well, we do have the MBA doctor up there to assist, but he's not a neurosurgeon."

"And how about the time delay? Even with the haptic receptors that give you the sensation of touching what the VR robot is touching, it's still not the same as being there, is it?"

"Surprisingly close." I found myself nodding. It was amazing how real the VR unit made it feel. "We've been doing robotic surgery over remote data connections since the late twentieth century, and the equipment has gotten better and better. Don't ask me how it all works, but the feel is pretty damn close to the real thing, and the time delay…well, there is none, at least not enough of one for me to notice. I'm still not sure how they pulled that off."

"Atomic wormhole generator," Green said matter-of-factly.

Green was only about 10 or 15 years younger than me, but still closer to this new generation that takes for granted technology that was only Science Fiction when I was a kid. I'd heard about wormholes, of course. But I hadn't the foggiest how they actually worked.

"They've had it up and running for a few years now."

"Whatever." I shrugged. "Anyway, it's the same as sitting in a room next to the patient, once I'm in that chair. Like I said, nothing heroic."

Green seemed to be paying more attention to my body language than my words, his gaze darting between my eyes and my hands, with frequent breaks to glance at the old-fashioned analog clock sitting on the side of my desk. "Uh-huh, uh-huh. And how many times have you used it?"

His tics were getting hard to ignore and I found myself wondering how he'd advanced this far in such a public career. I tried to politely ignore it. "For an actual operation, just the once. We train on it for an hour a week to keep up to speed."

"We?"

I nodded. "They picked ten of us at start-up: me and my partner, Doug Wiley, along with two each from vascular surgery, cardiothoracic surgery, general surgery and orthopedics. They figured the MBA doctor could handle anything else. As far as I know, I'm the only one who's actually used the damn thing. Not very cost effective."

He wiped a bead of sweat off his temple with the back of his hand. "Doctor Baker probably wouldn't have agreed with you on that one."

Sue Baker, the geological engineer I'd operated on, had served just 2 weeks at Moon Base Alpha when they rolled her into the MBA operating theater under my robot counterpart. A CT scan confirmed the subdural hematoma that was compressing her brain, and an hour later I was fusing the cranium shut with a laser bone-welder after draining the offending pool of blood.

I smiled. "I suppose you're right. Wonder if she's still up there?" I'd released her for duty about 3 weeks after the surgery. Follow-up scans and neuropsych testing were normal, except for the brief period of pre and post-traumatic amnesia that never cleared enough for her to remember what had happened.

"Didn't you hear?" Green's expression turned somber.

"What?"

"She died last week," he said, stone-faced.

I felt a twinge in my gut. No matter how hard you try to maintain objectivity in dealing with your patients, you still become emotionally attached to them at some level. "So that's why you're here."

Another glance at the clock. "Well, yes, but only because it's stirred up interest in the robot again; made people think about the remarkable thing you did."

My feet were starting to go to sleep. I shifted in my chair. "How'd she die?"

"Air-lock failure. She was out at the Geo-Survey Station."

"Her partner, too?"

Green shook his head. "According to the wire report that came into my office, he was out collecting samples when the airlock failed."

"Don't they have like a million fail-safes for that system?"

"Supposed to. I guess they didn't work. They said it was a meteorite storm; one of them penetrated the hull of the building or something."

The numbness was creeping up my legs and I shifted again, with difficulty, my fingers now feeling the same burning tingle that had been plaguing my legs.

Green's ruby complexion had washed out and he was fidgeting with his collar, awkwardly trying to spread it with his left hand, get some air. "You feel it, don't you?"

I squinted up at him, studied the pained expression on his face. No wonder the sonafabitch had been so damn nervous. I reached into the pen holder on my desk and pulled out a brushed stainless steel pen, longer and flatter than the others; started to fumble with it as I tried to make some sense of all this.

He watched my hands as I twisted the pen clumsily between my fingers. "I am *so* sorry. I wish there had been another way. I…I really do."

The bastard actually sounded sincere, but my jaw clenched tighter with every word he spewed, and I had to fight to keep my focus. "What the hell did you do to me?"

He pressed his teeth against his lower lip. "It's a neurotoxin. See, when you shook my hand…" He carefully peeled a clear form-fitting liner off his right index finger, turned it inside out, then dangled it gingerly in front of me. "They said it would work quickly." He dropped the contaminated liner into a plastic bag he'd pulled out of his jacket pocket, and sealed it in. "How long ago did you start to feel it?"

"Just a few minutes." I went back to fidgeting with the pen.

"Look, you've only got about another twenty minutes or so before all your muscles freeze up. Your lungs will stop working a few minutes after that." He paused and took a deep breath. "Unless…" He pulled a syringe out of his pocket. It was filled with a dark orange liquid. "We get this antidote into you."

"Then *do* it," I snarled.

He hesitated briefly. "No. I didn't come this far just to wimp out."

"What do you want from me?"

"I need access to those files, the records of your operation on Susan Baker. And the back-ups, of course. Once I erase them, you can have *this*." He waved the syringe in front of me. His hair was damp with sweat, but his resolve was undeniable.

I wasn't ready to die. Not yet. "Computer: On screen."

The image popped up in front of me again. I gave it my access code and called up the file. "Dual image," I barked at the machine. A duplicate view came up facing Green. "Allow secondary access."

Green puffed out a wisp of air. "Smart man." He reached for the screen and entered the sequence of commands needed to wipe the files; did it with the alacrity of someone who knew his way around computer code.

"And the back-ups?"

I could feel my breaths getting shallower, and the muscles in my arms seemed to be weighing down my hands rather than guiding them. I accessed the automated back-up system, the one that copied active records to the patient's permanent medical record file, as quickly as possible and turned control over to Green. He was remarkably adept at his job, at least the computer part of it. He seemed considerably less comfortable with the hit-man part.

"Done," he announced. "Screen off."

I mimicked his command and the image vanished from over the desk.

"Quickly," I said, motioning toward the jacket pocket where he'd stashed the syringe. "I don't...have...much time." The breaths were coming faster now.

Green pulled the syringe out slowly with one hand and mopped his brow with the other. "I don't know," he said, eyes fixed on the orange liquid.

"You got...what you came for. You really...want a murder rap...on your hands." I laboriously stretched a hand toward his.

He pulled back. I could feel his sense of humanity giving way to self-preservation. As he started to draw his hand back, I swung my arm up as straight as possible while activating the beam on the laser scalpel pen I'd been fiddling with. I'd already adjusted it to maximum strength, but I'd still have to get close; this thing was made for small depth cuts.

Firing up the laser required only a flick of the activating switch, a task I wasn't sure my fingers could manage in their current state, but they fluidly completed the pattern of movement that had been ingrained by years of repetition. The cutting beam tracked across the back of his wrist, searing down to the bone and detaching the muscles that had been keeping his wrist extended.

The syringe dropped from his weakened grip, landing on my desk pad with a sound that was overwhelmed by an agonizing scream. He grabbed at the wound with his unblemished left hand as he recoiled back against his chair, then threw himself forward, writhing in pain from the searing flesh.

I had known how he would respond, and prepared myself to strike. Every fiber of muscle that remained at my disposal contracted in agony as I lunged at him, thrusting the laser blade forward as high as my hand would take it. The poorly controlled motion made contact as it arced over his neck, striking a direct slice across the carotid artery. I fell against the desk just before his blood splattered across the polished dark oak surface in front of me.

His body writhed violently, but only for a few more seconds before he fell lifelessly into his chair. I mustered the strength to pull myself back just enough to clear the emanation point for my computer screen, so I could call for help. But before I could even activate it, Doris had burst into the room and a new round of screams began.

I couldn't lift my head. "Shut up…and give me…the damn shot," I said, motioning to the syringe with my eyes.

"Oh, God. Oh, God." Doris was a retired O.R. nurse who'd been working my front desk for the past few years. She was accustomed to blood and not easily rattled, but this was too much even for her. I wasn't sure she'd heard me between her screams.

She grabbed the syringe and palpated for a vein, unsuccessfully. I guess I looked even worse than I felt, and I felt like death warmed over. She plunged the needle directly into my carotid artery.

*

The morning sun warmed my eyelids, but the light was painful as I squinted, struggling to interpret the foreign environment that surrounded me.

"Well, good morning, sleepy head." I knew it was Doris before I turned toward the raspy, but comforting voice. She was the closest thing I had left to family.

"Where the hell am I?"

"You're welcome," she huffed.

Obviously, I should have been thanking her, but for the life of me I couldn't remember why.

She scowled.

I looked into her eyes and it started to come back to me. "The antidote?"

"Right in the carotid." She made a stabbing motion. "I always wanted to do that."

"In general? Or just to me?"

That got a laugh. I loved her laugh. Hadn't heard it much these past few years. She missed the excitement of the operating room, but the days of grinding out long hours on her feet had passed her by, with the help of a back injury that had almost made her my patient.

"Thanks," I said.

"You're welcome."

The door swung open and Josh Coggins, a close neurologist friend of mine, strolled in. "Well, look who's up."

"You make it sound like I've been in a coma, or something."

He looked toward Doris, then back at me. "Something." He studied the chart in his hand briefly. "You were damn near dead. If it wasn't for Doris, here…"

"She's already told me how heroic she was."

She was close enough to give me a smack on the arm.

"Yeah, well, she *was*. Still haven't figured out exactly what that toxin was, but it was nasty. You were barely breathing when the Code Blue team got there. Good thing your office is in the hospital. If we hadn't gotten you intubated and on the vent…"

"Intubated!" It hadn't occurred to me that I'd been through anything more than an unusually sound night's sleep. "How long?"

"We kept you sedated for a day or so, on the vent; decannulated you last night. You don't remember?"

Not a clue.

"I guess all the drugs haven't worn off yet. Probably still a bit foggy."

I was.

"Let's see how you're doing." He ran me through a quick neurological exam: cognition, cranial nerves, strength, reflexes, sensation. "Damn near back to normal."

"Good," I said.

I popped up out of bed and the room started spinning. Josh and Doris each grabbed an arm and laid me back down.

"Damn *near*, I said. It's going to take a little time, even for a *neurosurgeon*," Josh said with a snicker. "We'll get you some physical therapy today. My guess is you'll be out of here in a day or two."

I wanted to argue, but my body told me he was right. "Thanks, Josh."

He nodded and left the room.

I turned to Doris. "That asshole, Green…I killed him, didn't I?" It made me sick to my stomach to realize that I was actually capable of doing something like that.

Her face flushed and she looked at the floor. "Better him that you," she muttered. "The cops aren't pressing charges, but they want to talk to you as soon as you're up to it. Do you know why he did that to you?"

"Yeah."

She waited for me to elaborate. I wasn't sure if telling her more would help protect her or just get her in deeper. Before I could answer, she spoke again.

"Strangest thing—not two hours after the paramedics left, I got a call from administration saying the MBA medical office was requesting the files on Susan Baker. That's what Green was there to interview you about, right?"

"Yeah."

"I wasn't sure what to do, but when the boss calls…I tried to forward the files to them, but the records were gone, wiped clean."

I let out a bated breath. "You did the right thing, Doris." At least Green's men had confirmation that he'd completed his job before I killed him. They'd leave us alone now. Unless they thought Doris was covering up. Unless they thought I knew something that I did not.

"Doris, call Jimmy Valenti for me, would you? Tell him I need to speak to him ASAP." I'd first met Jimmy when he was a patient of mine—dug a bullet out of his back that had lodged a few millimeters from the spine. He was a beat cop back then, but had worked his way up the ladder to homicide detective. He always had some great stories; I never thought I'd be one of them.

"No need. He's outside the room, waiting. I told him I'd call when you were up to talking, but he said he didn't want you wandering off before he had a chance to see you, and that if he knew *you*, you'd be out of here as soon as your eyes opened."

That brought a smile to my face. "Would have, if my feet had let me."

"You want me to go get him?"

I nodded. "And wait outside for me, would you? Don't go back to the office, don't go home."

She studied my face. "What the hell's going on, Roger?"

"I just want you here with me tonight."

She laughed. "Right."

"Just do it, Doris," I said.

She looked at me and stopped laughing, then nodded and went out to get Valenti.

*

Jimmy Valenti looked more like a mob boss than your typical cop: a full head of dark slicked-back hair streaked with just enough gray to exude an air of authority, strongly chiseled facial features marred only by a barely perceptible angle where the bridge of his nose had been broken, and dark brown eyes that seemed to never blink. His nails were always manicured and I'd never seen him in anything but a fine silk suit accented with a custom tailored art-deco tie tucked under a perfect V-point collar. A bulge under his left breast pocket completed the persona.

He walked in and stopped at the foot of my bed. "You look like hell."

"Thanks," I said. "*You* look like a male model."

He turned his arms out to the sides and struck a pose.

I couldn't help a wisp of a laugh from sneaking past my lips. "Take a seat. You make me feel like I'm a patient or something."

Valenti pulled a chair up beside me and sat. "What the hell happened?"

I told him every detail I could remember.

"All that just to wipe out the Baker files? Did he make a flash copy?"

"That's the weird thing. He didn't even look at them or ask me anything about them, except how to access them. Then he wiped the primary *and* the back-up."

"Only one back-up?"

"It's a hospital, not the Securities and Exchange Commission. Memory is money. Besides, the more copies you make, the harder it is to stay in compliance with all the health-care privacy laws. One active file, one in the permanent medical record—that's the standard."

"What kind of reporter wipes the files before he even looks at them?"

"Obviously, he was after something besides the story," I said.

"You *think*?" Jimmy fired back.

I glared.

"Question is," he said, "what was he *really* after, and who was he after it for? Reporters don't usually go around killing their sources."

"Not to mention the fact that you don't just pop into your local pharmacy to pick up an exotic neurotoxin, and I doubt it was something he'd whipped up in his kitchen."

Valenti arched an eyebrow. "Somebody's got something on him; used him because they knew he could get access to you without raising any red flags. Only he was so sloppy that he not only raised one, he ran it up the flag pole where it could flap in the breeze."

"Makes sense," I said. "He was more nervous than *I* was in there."

"That's why they call me a detective." Valenti gave me a pat on the arm and stood. "I'll see what I can dig up on him. I'll run his financials, first. It's almost always about money."

"Thanks, Jimmy."

He gave me a nod. "In the meantime, I'll assign a couple of guys to keep an eye on you and Doris."

"You think that's really necessary? Once the word's out that the files are gone, they won't have any reason to bother with me." I had my own doubts, but hoped Jimmy would quell the thought, assure me there was no way in hell they'd come after me again.

"Probably right, but it can't hurt. Indulge me. At least till we figure out what's going on here."

"Let me know what you find."

"You'll be the first," he said, then walked out the door.

*

I talked my way out of the hospital that afternoon, with Doris's help. She assured Josh she'd keep me out of trouble.

My condo was on the eleventh floor of Stolle Towers, just a stone's throw from the hospital. I'd been in the same place for nearly 30 years. It gave me easy access to the hospital when I was on call, and the spectacular view of the city, with the sun setting behind the rolling mountains in the distance, made it a great bachelor pad, in its day. In *my* day.

Doris, only a few months my junior, had been one of my early conquests. Not the best idea when you have to work together. The relationship had been steamy, but short, followed by over a decade of days in the surgical suite without having her assigned to any of my cases. And she was the best damn O.R. nurse in the hospital.

In time, the ice between us thawed, and whatever it was that had drawn us together in the first place turned out to be more than her smoking-hot body and my prestigious position. Once we could see each other without throwing verbal daggers, we realized that we still enjoyed each other's company. We didn't let it happen too often.

A few years back, Doris opted for retirement, but the boredom was eating away at her. She called one day to ask if I knew of anything that could keep her busy around the hospital, without having to put her arthritic back through long O.R. shifts. *Fate:* My secretary had just gone on maternity leave. That was 7 years ago.

Valenti's men drove us home, with a short stop at Doris's place to pick up some clothes and whatever else it is that a woman has to pick up to survive a couple of nights at bachelor pad. Or in this case, a safe house. One of them led us in after I deactivated the door lock, the other trailed behind with the suitcase.

"Where would you like this, ma'am?"

Doris looked at me.

"Second door on the right." I pointed down the hallway.

Her left eyebrow crept up as she cocked her head without diverting her gaze from me.

I turned my hands out, pleading innocence. "It's the guest room."

She kept me fixed with her dark green eyes and said with a tone of disappointment, "I remember."

A tingle crawled up my spine. "Really?"

She grinned. "Just pulling your chain." Then to the cop with the suitcase, who'd been watching our exchange, she repeated my directions: "Second door on the right."

Like I really needed this right now.

I watched the burly man in blue toting her bag down the hall like it was filled with feathers, and sighed. "Quite the pair, huh? Between your back and my pathetic muscles…" I remembered feeling a little different the last time Doris was here. Twenty-seven years… *Man.*

She took my arm. "Not so pathetic before that jerk got to you." Her hand gave my floppy biceps a gentle squeeze. "You'll be back to yourself in no time, glad to see me go."

I wouldn't, I realized to my surprise. Be glad to see her go, that is.

The guard had returned. "We'll be taking shifts. One of us will be right outside the door, if you need us."

"Thanks."

When the door closed, Doris and I were alone in what should have been awkward silence. But it wasn't awkward, at least not for me. From the gentle grasp that caressed my arm, not for her either. Working closely in the office, we'd been alone many times in recent years, but not like this. If this all had happened a few years ago, maybe even a few months ago, I was sure I would have felt differently.

She helped me onto the sofa.

"Can I get you something to drink?"

I nodded. "Some iced water would be great."

I heard the tap go on.

"From the cooler," I called out. "Next to the fridge."

"That's new. A little elitist, don't you think?"

"I couldn't stand the metallic tap water in this place anymore. I finally gave in a couple years ago and arranged to have Celestial Springs deliver a fresh five-gallon container every week."

Doris started to hum, then sang: "Ooh, ahh, nectar of the moon and the brightest stars," to the tune of the company's jingle, one of those irritating little ditties you can't get out of your head for hours.

She knew where the glasses were. Men don't rearrange their kitchens too often. She returned with two glasses of water, and sat. "You really think they'll come after us?"

"Nah. They got what they want." I wasn't very convincing.

"Then why that look?" She knew me too well.

I let out a huff. "Why *now*? Why, after all these years, do they want those files erased so desperately that they're willing to kill for it, willing to risk a high profile murder?"

"Sorry to break it to you, Roger, but you're only famous in your own mind." She tried not to laugh.

"Not *me*." There was a time I'd have shot daggers at her with my eyes for that one, but I guess I'd asked for it. I let the opportunity slide. "*Green.* He may not have been the host of *Good Morning America*, but he was enough of a celebrity that they had to know that stunt of his would draw a lot of press if it didn't go perfectly. Why would they risk that?"

"They must have thought it was their best shot of wiping out those files."

"And me," I had to admit, no matter how much I'd tried to convince myself otherwise. "They don't want to take the chance that I've seen whatever it is they don't want anyone to know about."

"Oh, God, Roger."

"Yeah." I tried to stand, but flopped right back into the deep sofa cushion. "Crap. Help me up, would you?"

She started to reach out for me.

"No. Wait. I don't want you to hurt your back. Just bring my Net-pad. It's on my nightstand."

She stood.

"And…ah, what the hell. In my closet, behind the underwear drawer, there's a dead-panel."

"A what?"

"A dead-panel: a piece of plywood covering a hidden compartment. Just press both bottom corners at the same time and the panel will pop out. There's a blue bubble-chip in there. Bring it to me, would you?"

"Your underwear drawer?"

I nodded.

"Good spot. Who the hell would go near *that*?" I didn't need to see her face to know what that smug grin looked like as she headed for the bedroom.

"Just the blue chip. Don't touch anything else."

"Yes, doctor," she called back with a salute.

I admired the lilt in her step as she walked away. Doris was one of those rare women whose looks only changed for the better as the years went by. She'd let the gray sneak gracefully into her auburn tresses, but never let her body lose its tone. I hadn't been in this apartment with a woman my age since…well, since I wasn't anywhere near this age.

She returned with the chip and my Net-pad. I pushed a button between my seat cushion and right armrest, and a tray table popped up from the arm of the sofa. I flipped it down in front of me and sat the pad on it, then pushed the power button. A 14 inch viewscreen appeared over the pad. I inserted the bubble-chip into the port of the pad, and with a few swipes of my very tired finger, found the file I wanted.

Doris had sat back down next to me, watching the proceedings. "You copied it? Jesus, Roger. If anybody found out about that, you could lose your license."

"Better than my life. Besides, *you're* not going to tell anybody, are you?" I didn't need to wait for an answer. "So who's going to know?"

"You do this with all your files?"

"Hell, no. But this one…it was the first robotic surgery case on the moon, for God's sake. And *I* did it. How cool is that?"

"So how many times have you massaged your ego looking at this thing?"

"Including now? Once. Who the hell wants to watch themselves operate? I just wanted to know that I had it."

For the next hour, I reviewed the data, studied the scans of the surgical area, even watched the surgery itself. Routine case from start to finish. Except for the circumstances.

*

The phone chimed softly, innocuous enough to almost ignore at first, but the sound went on endlessly until I finally acquiesced and tapped the receiver on. "Bennett," I forced out through my parched throat.

"Roger?" It was Valenti's voice. "That you, Roger?"

"Jesus, Jimmy. You realize what time it is?"

"Yeah. Do you?"

I glanced over at the clock on my nightstand: nine-twenty AM. "God, it feels like the middle of the night. I haven't slept this late in… well, I can't remember when."

"Yesterday," Valenti said. "In the hospital. There's a reason they told you to stay there."

I tried to shake the cobwebs out of my head. "I'm fine."

"Yeah, well don't plan any big outings. Look, I ran the search on Green. The guy's squeaky clean, a friggin boy scout. Literally. Got his Eagle Scout during high school, graduated from MIT with a degree in computer sciences…"

"Guess that explains how he found his way around my office file system so easily."

"Which somebody was obviously counting on. Only we'll never figure out *who,* from these records."

"MIT seems like an odd choice of schools for someone who wanted to get into the broadcast news industry."

"Sounds like he fell into it—started video blogging the local news in Boston as a school project during his senior year and was so damn good at it that NetNet News offered him a job out here right after graduation."

"Great. So why the hell was he willing to risk it all to come after me?"

"No idea. I ran his financials. Whoever was pulling his strings wasn't doing it with money, at least not that I can find, but there's got to be something. It just doesn't make any sense. I'm going to do a little more digging. Meanwhile, you and Doris sit tight. I don't like the way this is shaking out."

I scratched my head and yawned. "Right."

"Good, I'll…"

"Say, Jimmy. What do you know about that accident at MBA last week, the one Baker got killed in. I ran a Web search, and there's some talk about a freak meteor shower that punctured the mobile lab she was working in, but the timing seems a little too coincidental, don't you think?"

"Baker? Susan Baker? The one whose file Green was after?"

"One and the same."

"Shit. I don't usually pay any attention to what goes on up there, but I guarantee you that ain't no coincidence. I've got a buddy down at NASA who might be able to get me some details. I'll give him a buzz."

No sooner had he hung up than I heard a gentle tap on my door.

"S'open," I yelled.

The door swung slowly into the room, and Doris's face edged out from around the other side. "You decent?"

"I try to be."

She didn't laugh, but by then she'd seen that I was safely under the covers. She walked in and sat on the side of my bed.

"How did you sleep?"

"Too well."

"I didn't want to wake you. Coffee's up and I scrambled some eggs. You still like them scrambled?"

"My arteries don't, but I do." I looked forward to a real breakfast. I never bothered making one for myself. "I'll be in in a minute."

She got up and turned to go.

"Thanks," I said.

She looked back over her shoulder and smiled. "Can you manage OK?"

"No problem." I hoped. But either way, I wasn't about to ask her to dress me, for God's sake.

"Don't be a hero. I don't want to be picking your broken ass up off the floor."

"I'm *OK.*"

She walked out and closed the door.

I sat up on the side of the bed and waited a good minute or so before the room stopped spinning. After that I really *was* OK, I was pleased and somewhat surprised to find out. I felt much stronger than I had yesterday. I treated myself to a quick shower, then headed out to the kitchen following the aroma of freshly brewed hazelnut coffee.

"Damn, that smells good." I started to reach out to pour myself a cup, but Doris blocked my hand.

"Just sit," she said. "A little bit of Half and Half, no sugar. Right?"

I gave her a nod. "Thanks."

She brought over a mug of steaming hot coffee and a plate of eggs with some toast.

"Where did you get…"

"I sent one of the boys to the market downstairs."

"The boys?"

"It was change of shift for our personal body guard out there." She motioned toward the door. "I figured since his relief was here, I could send him down for supplies."

"I'm sure he loved that."

"He's trained not to complain. Too bad I can't teach *you* that."

I started to protest, then realized if I did complain I'd already lost that argument. I blew some of the rising steam off the top of the mug and took a sip. "Umm, that's perfect. Thanks."

"So was that Valenti on the phone?"

I had a mouthful of eggs, so I responded with a nod.

"What did he find out?"

I swallowed and cleared my throat. "Not much. Seems Green was a perfect little angel."

"Until he tried to kill you."

"Yeah. Till that."

"So Valenti still has no idea why he came after you? Why he wanted those files?"

I took a long draw from my mug, holding it with both hands, then set it down. "Nope, not a clue. I asked him to dig up what he could on that accident up at MBA last week. Maybe he'll find something there."

"I tried to read about it on the news threads this morning," Doris said, "but there's practically nothing."

"Yeah, me too. I poked around the Web last night before I turned in. They want everyone down here to think it's like Disney World up there so their funding doesn't get slashed, but Jimmy will find out what happened. Meanwhile," I started looking around for my Net-pad, "have you seen my…"

"Under the paper. Figured you might want it. Didn't notice when you tossed the newspaper down, huh?"

I hadn't. I treasured a real paper to keep me company at breakfast. Enough people still did, that they'd kept up home delivery for the Sunday Times, and I had a system for nursing it through the week, one section at a time. Today it was Sports, which I'd carried in from the hallway accent table on my way from the bedroom. I lifted the crinkly pages and my Net-pad was right where she said it would be.

I powered up and started a Web search on Green. No point in checking for financials, business links—that sort of thing. If Jimmy couldn't find any dirt through police channels, I certainly wouldn't find anything on the Web. But I figured I might be able to learn a little about him through his past. There had to be something to connect him to Susan Baker or NASA. Maybe one of his MIT classmates.

An hour of dickering with my Net-pad accomplished nothing but taxing my meager endurance. I logged off, poured myself another cup of coffee while Doris showered, and settled for some not-quite-up-to-the-minute stories in the Times' Sports pages.

*

It was barely noon, but Doris and I had already started to go stir-crazy, something that happens ridiculously fast as soon as you know can't wander out of whatever place you're in, no matter how cozy or how good the company. The ring tone announcing Jimmy's call couldn't have come soon enough.

I grabbed for the phone. "Jimmy?"

"You're not going to believe this, man."

Ten minutes later, he was in my living room, sitting on the taupe leather chair across from me and Doris. "The guy who called in Baker's 'accident,'" he air-mimed the quote marks, "was the same clown who pulled her to safety after the first 'accident' a few years back."

"Is that really so strange?" Doris asked. "I mean, there aren't all that many people up there, are there?"

"I guess not, especially since they were the team assigned to GSS2 on both occasions, but still…"

"GSS2?"

"Oh, yeah. Sorry." Jimmy placed his Net-pad on the coffee table between us and activated it. A screen popped up, with matching images visible from either side.

"Computer," Jimmy barked, "expand display to twenty inches."

The computer complied.

"Access file: MBA base map."

A map of the nascent MBA village came up, showing an overhead of the original dome in the center of the complex, and a series of buildings that looked to be haphazardly arranged around it.

"Zoom out."

The size of the village slowly shrank as the moonscape came into view. It had shrunk to no more than a couple of centimeters in size by the time some of the far side of the moon began to appear on the left side of the screen.

Jimmy let it pan out a few more seconds, then called out, "Stop zoom. Highlight geo-survey plan."

A rectangular grid appeared over a portion the shadowy far side of the moon just past the border of light, forming a row of four equally sized squares, each grid-marked in a different color: white, green, blue, and red. Judging from the miniscule size the village now appeared, I figured the area had to be hundreds of miles across.

Jimmy motioned toward the image. "This is the most recent schematic available for the Geological Survey Project of the MBA facility. They laid out a standard grid pattern to begin surveying for minerals, ores…anything that might prove to be of use or interest on the far side where we've never been able to get much information before. Each color represents a ten thousand square mile geo-survey area.

"Computer: Zoom to Geo-Survey Area 2; factor of three."

The image closed in on the green area, revealing a ten-by-ten box grid.

"Each geo-survey area is divided into a hundred segments, and each segment is ten-by-ten miles. Since the original colony only consisted of twelve people, they'd only brought two geologists along, Sue Baker and David Parsons. The green area was the closest to MBA, so they started there—Geo-Survey Area Two, and worked out of a mobile laboratory they call a Geo-Survey Station."

"GSS 2," I said. "Where Baker was injured."

"And later killed," Jimmy added. "They have a lab set up on site so they don't have to waste time and resources travelling back and forth to the main base. But calling it mobile…well, that's a bit generous. It moves about a half a mile per day at top speed. They work one square at a time, going out on surveys during the day and returning to the GSS lab to analyze the samples each evening. They sleep there, then start over the next day, covering each 10 mile segment in about a month."

"At that rate," I calculated, "it'll take about eight years to cover each Geo-Survey Area."

"They've been gradually adding more units. They have four teams operating up there now."

"So the two people on each team are marooned out there for months at a time?" Doris asked. "I think I would have killed her myself after being cooped up all that time in such close quarters."

Jimmy laughed. "Careful, Roger. Sounds like I'd better get at least *one* of you out of this apartment soon."

I eyed Doris over. "I think I can take her."

She sneered back. "Just keep thinking that."

Jimmy cleared his throat. "Anyhow, the way I understand it, they shut the lab down for 2 days every week so the team can return to base for some R&R. They have a hopper that can get them back in a couple of hours."

"OK," Doris said. "So he had the opportunity and the means to kill her, but other than generally getting on each other's nerves, what motive?"

I smiled. "All that TV is finally paying off."

She answered me with an elbow.

"She's right, actually," Jimmy said. "I mean, that first crew was hard-core NASA—the kind who can put up with misery that would break any of *us* in the first hour. And if this was murder, it was no crime of passion. It was well-planned, calculated."

I looked up from the screen I'd been staring at. "You got some details?"

"Oh, yeah. From the official report, it seems Parsons came back from a day of collecting core samples to find that the GSS had depressurized. He'd last checked in with Baker twenty-five minutes earlier—standard thirty minute check-point contact—and she'd just gotten back to the lab to run her samples. He didn't realize anything was wrong until he entered the pressure hatch and tried to equalize. That's when he found out there was nothing *to* equalize; the air was gone."

"Sure," I muttered. "And if you believe that I've got some waterfront property in the Mojave to sell you."

"That's what's in the report," Jimmy said. "And whatever caused the GSS to depressurize must have happened pretty quickly. Baker didn't even have time to get back into her suit. She was sprawled out on the floor when Parsons found her; ugly sight, from the report. I'll spare you the details."

"Thanks," Doris said. "I've had enough of the blood and guts stuff for a while."

"He sealed the place up, and they sent a team to investigate. They found a couple dozen holes in the hull, each about a centimeter across; figured it must have been a meteorite shower."

"Parsons have a gun with him?" I asked.

"A better question," Doris said, "is why did Baker just sit there waiting to die? I mean, even with all those holes in the wall, you'd think she would have had enough time to get to the hatch and try to get into her suit."

"Guess she was trying to seal the leaks."

"That's something a man with a hero complex would do. A smart woman like her would get the suit on first, and *then* seal the holes."

"Maybe the air lock was breached or just not working."

"Or maybe," I said, "just conveniently jammed shut."

"Something I doubt we'll ever know," Jimmy said.

He shut off the Net-pad.

The three of us sat there numbly, trying to figure out how all this was connected. At least that's what I was thinking. Baker's head injury a few years back, Green's attack in my office, the so-called meteorite accident—it had to all tie together somehow. But sitting there ruminating on it wasn't helping. The more I thought, the less it made sense.

I looked up at Jimmy. "Still nothing on Green?"

He shook his head in disgust. "Squeaky clean. The sonofabitch doesn't even cheat on his taxes. No payola, no reason for him to try and turn you into dog food, as far as I can tell. Did you ever date his sister?"

I glared. "Cute, Jimmy."

He shrugged. "If the condom fits…"

Doris laughed. "I guess he *does* know you pretty well, Roger."

"He's *joking*. I'm not that guy anymore. Tell her you're joking, Jimmy."

He paused, right eyebrow creeping slowly up his forehead.

"Jimmy!"

He broke into laughter. "If he ever *was* like that, I'd love to have seen him in action, but…"

"No," Doris interrupted. "You wouldn't have. It was not a pretty sight."

Jimmy shrugged. "Anyhow, the Roger I know is anything *but* a lady's man."

"You're not helping here, Jimmy."

"Well, compared to the crowd I hang out with, you're vanilla, man."

I couldn't argue. I'd seen his crowd. "How about Parsons? Any leads there?"

Jimmy's face soured. "Got my boys working on it. I'll let you know." He picked up his Net-pad and slipped it into his pocket as he got up to leave. "Well, I'll leave you two to…whatever it is you do." He gave us a wink.

"We *talk*," Doris said. "Occasionally."

He raised his eyebrows at me. "Good luck with that."

"I'll walk you out." I came up next to him and gave him a pat on the back. "Thanks for all this, Jimmy."

"You bet."

I opened the door and held it for him. The uniformed cop he'd stationed in the hallway stood vigilantly as Jimmy headed for the elevator. I watched it close and went back inside.

As I looked at Doris, a thought occurred to me. "Didn't your college room-mate have some highfalutin position over at NNN?" NetNet News had become the leading source of news for most major cities, carrying local streams in each city, followed by national and then international streams later each evening.

"Sherry Stein? Yeah. Director of Personnel, last I heard. God, I haven't talked to her in ages."

"No time like the present to renew an old friendship."

She crossed her arms over her chest. "Green? You really think she's going to risk her job to give me anything wrapped so tight that even Valenti can't get to it?"

"Don't ask her for that stuff. Just see if you can find out who he hung out with, maybe somebody we can talk to. I can't just keep sitting on my hands all day."

It took her a while to track Sherry down, but once she got her on the horn, she was talking so fast I felt like my head was spinning. She worked the conversation until I finally heard something of interest.

"Heard about it? I was *there*. ... Yeah, my *boss*." She rolled her eyes. "Uh-huh, the same Roger Bennett. ... I *know*. Can you believe it?... That was a long time ago. ... Oh, *I remember*." She tilted her head in my direction and I glanced away. "I didn't think I knew those kinds of words either. ... *No*, I'm not dating him! I told you, he's my *boss*."

I felt my teeth clenching as the conversation wore on.

"I know. He tried to *kill* Roger. Was he on drugs or something?... Really?" Her face squinched. "A puppy? Well, you'd never know it from the way he acted in our office." Her eyes widened. "You didn't. And I thought *my* taste in men was bad. ... Well, at least you cut it off before the *real* Tarin Green came out in him. What a lunatic." Then her shoulders dropped and the tone of her voice softened: "Just the same, it's really sad he had to die so young. Did he have any family?" ... "Really!" Her back arched. "God, what a jerk! So, I guess he never told you much about her, then, huh? ... A daughter up at Moon Base?" Doris gave me a nod. "That *is* a lot to be proud of. She still up there? ... Too bad. It's sad there's nobody here to go to his funeral. I mean, family, you know. ... Well, co-workers aren't quite the same thing, but I know what you mean. At the hospital, we're all kind of like family too." She turned away, but only after I'd noticed the serene smile on her face as she said it. "No problem. I know you must be busy. ... That'd be great. Maybe we can catch up over coffee, sometime. Give me a call when things settle down. ... You too. Bye."

She disconnected, then crossed her arms and gave me a smug stare.

All I could say was, "What *kind* of words?"

"Huh?"

"Oh, never mind. I probably deserved it, back then."

"Yes. You did."

"So what's this about a daughter?"

"Turns out Green had a daughter he didn't even know existed until a few years ago. She showed up at his office one day and just sprung it on him. Apparently, her mother never wanted her to know who he was, but she eventually figured it out and wanted to meet him."

"Surprised Jimmy didn't find out about that."

"Sherry said Green kept it on the QT; made her promise to not tell anybody. He was afraid it might damage his public image."

"And she kept all this from the cops? For that weasel?"

"Nah. They haven't even talked to her, and unless they're going to check out all his old girlfriends, I doubt they ever will. She was transferred to the Seattle office six months ago; hadn't talked to Green since they broke up last year."

"She actually dated that guy? What was he—forty, forty-five, maybe?"

"Must have been older than he looked. He had a daughter in NASA, didn't he?"

"Yeah, but still…"

"And you're implying…what?"

"He just seemed kind of young for someone Sherry's age." I knew I'd made a mistake as soon as the words escaped my lips, knowledge confirmed by the glare coming my way. "Not to say that Sherry's…that *you*…ah, hell, you know what I mean."

Her frown cracked.

"You enjoyed that, didn't you?"

"More than you know. I *know* how old I am, Roger. I'm comfortable in my skin. Can you say the same?"

I couldn't.

"Anyhow," she said, "the reason his daughter had chosen *then* to come see him was that she was getting ready to ship out to join the survey team at MBA; wanted to meet him before she left, in case…well, you know."

I did. And then it hit me. "Oh, shit."

"As in 'Oh shit, Doris. You're good.'"

"Yeah. *And* as in 'Oh shit, I know what happened.'"

Doris stared for a few seconds, then threw her hands out to her sides. "Well?"

"Leverage," I said. "Green may have been the perfect guy to send after my files—his computer background, a good cover story to get me alone in the office—but a guy like Green doesn't agree to do something like he did unless somebody's got some leverage to make him do it. Usually, it's money or some dirt—an affair or something, but Jimmy didn't find any of that. Because it was the daughter."

Doris stood motionless for a brief second before it sunk in. "Of course. They threatened to kill the daughter if he didn't do it. *That's* why he was so freaked out about Baker's death. He knew they'd really do it if he didn't destroy those files."

"And getting rid of *me* was obviously part of the package. This guy was no natural-born killer, he was almost as terrified as I was. They must have made getting rid of me part of the deal to save his daughter. A few more minutes and he'd have succeeded, too. He'd have come running out of the office calling for you to get help. It would have looked like I had a heart attack. Nobody would have thought to check me for some exotic toxin."

Doris looked a little ashen, obviously thinking back to the bloody sight she'd walked in on. She refocused with a subtle head shake. "So what the hell was in those files that they were willing to kill for."

"You mean, what did they *think* was in those files. I stared at those things until I was ready to pass out. There's nothing there."

So we were back where we started. Almost. We knew why Green was willing to come after me, but we still had no idea why they wanted the files, and what it all had to do with Baker's death.

I was going to have dinner delivered from the Italian joint down the street, but Doris insisted on cooking, saying it would relax her. I opened a bottle of Beaujolais, poured myself a glass and settled into the living room with the current edition of *Analog Science Fiction and Fact*, picking up in the middle of a novelette I'd already started.

Periodically pausing to give my glass a swirl until the last swig of wine was gone, I worked my way through a story about a Mars colony that gets established after an exploratory mission finds a water supply large enough to make a thing like that work. The twist was that support from Earth gets cut off due to budget constraints at NASA, and by the time trade is re-established many years later, the Martians have developed such a different set of morals that they don't want anyone from Earth coming to interfere with their lifestyle. Yeah, right—*Martians*.

As I ruminated on the thought, Doris announced that dinner was ready. The smell of French bread, just out of the oven, drew me into the room.

She'd dug out some fine china I'd forgotten that I'd inherited from my grandparents, dimmed the lights and put on a play list of my favorite classical music that started with Ravel's Bolero. A freshly poured glass of wine was in front of each setting, mixed greens decorated the salad plates, and a porcelain bowl of steaming Beef Stroganoff was placed in the center.

"Wow."

"Glad you approve. Now sit down before it all gets cold."

I hadn't realized that I'd been standing frozen in my tracks, staring at the table. "Oh, uh…" I looked over at her. She'd let down the fine auburn hair that I was accustomed to seeing tied up in a bun, and had obviously picked up some make-up when we'd stopped to get some of her things. It was hard to believe this was the same woman I'd barely noticed wandering around my office every day.

"Everything's lovely," I said, staring into her eyes.

"It's how it *tastes* that counts."

I leaned into her, put my hand behind her head and pressed my lips against the red gloss that covered hers.

It felt so right. To both of us. I could tell.

We separated slowly, and she cleared her throat. "Sit," she said. She brushed a strand of hair from her face. "I'll get some water."

She turned and went into the kitchen, returning a few seconds later with a small crystal pitcher in her hands. There was an awkward moment of silence as she poured a glass for each of us, then took her place across the table from me.

"Wanna talk about it?" I said.

She shook her head. "Just eat." She handed me the bread basket, then started in on her salad.

"It felt right, Doris."

"It did the last time, too."

I hadn't realized I'd hurt her so much when I'd broken it off those many years ago. It had been just another in a line of female trysts for me. I wondered how many other women had taken my actions for something more than they were meant to be. "You know I'm not that man anymore."

"Let's not spoil tonight, Roger."

I was surprised to discover that I had these feelings for the woman I'd worked with every day for the last few years. Maybe I didn't. Maybe it was just the circumstances. I decided she was right and changed the subject. "So you like Classical, too?"

"I do. Intricate, complex, and infused with feeling, yet so much less distracting than the stuff you hear on the satellite stations these days. It relaxes me."

"Mozart's still my favorite, all in all. But there's nothing like the Bolero." As I said it, I had no problem blaming Ravel for that kiss. Mercifully, the song ended, and the brilliance of a Vivaldi trumpet concerto changed the tide of the conversation.

She took a sip of wine, then put down her glass. "So, you're feeling better today?"

"Incredibly, yes. I guess it's true that a near-death experience heightens your senses. It hasn't felt this good to be alive in a long time." I scooped out a few serving spoons full of the Stroganoff and took a bite. "Oh. Wow. This is awesome," I sputtered as I chewed.

Doris laughed. "No manners, but thanks."

With the tension having simmered away and the after effects of wine helping free us from the worries of this bizarre set of circumstances that had quarantined us together under armed guard, we escaped into each other's company for the evening.

Under the mantle in the living room, a laser light show meant to pass for a fire served as a weak substitute for the real thing I'd experienced at my boyhood home in Vermont, but was soothing none-the-less. We settled onto the sofa across from the fireplace, maintaining a safe distance from one another, and popped up the leg rests. With a tap of the remote by my right hand, the 4 and a half by 8 feet picture of Monet's *Water Lilies* on the facing wall faded away, replaced by the all-to-familiar eagle logo of the Randall Corporation against a bright blue background, the iconic image that adorned the boot-up screen of nearly every home-theater media streamer in the country. Another tap brought up a menu of movies.

We studied the list together, glossing past romantic comedies that were sure to create awkward moments, and murder mysteries too close to our present reality, finally settling on a light-hearted fantasy, one of those sword and sorcerer things set in the Middle Ages. It had its desired effect of mindless escapism that I hoped would allow me the luxury of a sound night's sleep.

We said our casual good-nights, but knew casual was not in our future.

*

Jimmy was over bright and early the next morning, sitting across the breakfast table from us. We stared impatiently at the air over his Net-pad for a few seconds before the sixteen-by-nine inch screen image finally appeared, hovering just above the table.

He swiped a finger over some icons to pull up the file. "Ah, there it is. David Lionel Parsons. Born April 15…yada, yada, yada. Let's get to the good stuff." His eyes scanned the page. "There," he pointed. "Graduated from the University of Colorado School of Geological Studies with his doctorate. No big surprise in that, but look how he got there: two years at Emory School of Medicine, one of the top students in his class, then dropped out when he got a full scholarship to attend UC's geology program."

I'd been playing with my coffee mug, nudging it around by its handle as Jimmy talked, but that struck a chord. "Why the hell would he drop out of medical school to study rocks? He must have worked his ass off to get into Emory. It's hard to imagine giving that up for a little scholarship money."

"Maybe he couldn't hack it," Doris offered.

"Maybe," I said. Then, in my best raspy Don Corleone voice, added, "or maybe somebody made him an offer he couldn't refuse."

That got a weak chuckle out of Jimmy. Doris just glared. "You did *not* just say that."

I shrugged. "And yet, oddly enough, I meant it." I looked at Jimmy. "Where did that scholarship come from?"

He ran a finger down the screen, then stopped, tapping a finger in the air in front of the image. "The Randall Corporation."

"Randall? What the hell is a media company doing doling out college scholarships to geology students?"

"Actually, that part's not so strange," Doris said. "I saw a piece about Richard Randall on the Biography Channel a few months back. He may be a cut-throat businessman who clawed his way to the top of the entertainment industry, but he's got a philanthropic side too. He's an ardent supporter of the arts and education. They said he's put over a thousand kids through school with his scholarship program."

"And Parsons was one of the chosen ones, huh?"

"Apparently so."

"Which brings us back to: Why Parsons, and why geology?"

Jimmy looked up from the screen. "I think you're reading too much into this."

I pointed to the screen. "What else does it say about him?"

Jimmy looked back at the screen. "He had been working his way through med school as a research assistant with a Dr. Jeffrey Horsten…something."

"Horstengielfer?" I asked.

"Yeah, I guess that's how you pronounce it. You've heard of him?"

I had. "He did some of the foundation work on deep brain stimulation."

"Yeah," Jimmy was still looking at the file. "It's right here. Parsons got his name on a study they published in the *New England Journal of Medicine: The Effect of Deep Brain Stimulation of the Hippocampus in Primates.* Sounds thrilling."

"I'll be damned," I muttered.

Doris and Jimmy looked at me.

"What?" Jimmy asked. "What does frying monkey brains have to do with anything?"

"The hippocampus—it's where short term memory is processed." I looked at Doris.

"Hmm…now, where did I put that bubble-chip we were using yesterday?"

"Cute, Roger."

I smiled. "Would you mind digging it out for me again?"

"From your underwear drawer? Oh, I was *so* hoping you would ask."

I ignored her smirk and reached for my Net-pad, still sitting on the kitchen table. Jimmy's wouldn't have the software to run my medical files.

Doris was back in a few minutes with the chip. I put it in and pulled up the Baker file.

"Holy shit, Roger. Is that what I think it is?"

I nodded.

"You copied a patient file and brought it home? Isn't that illegal?"

I shrugged.

"You do realize I'm a cop, right?"

"What are you going to do, arrest me?"

"Arrest you? I could kiss you."

"That would be very bad for our friendship."

I pulled up the high resolution CAT scan of Sue Baker's brain, rotated it until I got to the back of her head and zoomed in on the occipital protuberance, that little bony lump in the back of the skull that nobody realizes they've got until they bang the back of their head on something, then swear that the lump they're feeling was never there before.

"Son of a bitch," I said. "All this time, I've been looking in the wrong place." I pointed to a miniscule black dot on the skull, only visible at high magnification. "It's a screw hole, about two millimeters in diameter." I rotated to the front of the skull and found two similar holes, one above each eye in the supraorbital ridge. A little more searching revealed the fourth and fifth marks on either side of the skull, just behind the ears. "The five anchor points for a stereotactic frame."

"I hate it when he gets like this," Doris said. "Speak English, Roger. I think I know what you're getting at, but Jimmy's eyes are starting to glaze over."

"Oh, uh…sorry. A stereotactic frame—it's sort of like a helmet, but just the framework. An oval titanium strip that's screwed into the skull at those five points, with a semicircular strip attached to it that goes over the top of the head from the nose to that occipital bump I showed you. The last piece is another semicircle of titanium attached by joints to the oval, just over the ears. It pivots 180 degrees from the front of the head to the back." I illustrated the motion with my hands around my head.

"It's computer guided," Doris added, "so that it creates a template to localize any precise point in the brain. They use it to focus radiation treatments."

"And," I added, "to guide the nanowires used for deep brain stimulation. Parsons could have used it to wipe Baker's memory; make her forget something she had recently seen."

"Then why whack her over the head?" Valenti asked. "Isn't that what you operated on her for?"

"Yeah. A subdural hematoma caused by a blow to the head. But think about it: if she just showed up at the base with sudden severe memory loss and no trauma to blame it on, it would have raised a few eyebrows. Parsons had to come up with a plausible reason. A blow to the head was the perfect cover, though I've got to admit, I always wondered why her memory loss was so severe. I mean, you never know with a head injury, but it sure seemed out of proportion to the trauma." I shook my head. "No wonder."

Doris bit her lip in thought. "But wouldn't he have needed a scanner to do what he did?"

"Nah. I mean, sure, you need it for precise localization of the nanowire in the brain, but all he had to do was get close to the hippocampus and give it a good jolt. It didn't really matter if he created a little havoc in the surrounding brain tissue. He knew he'd be able to blame it on the concussion."

"The real question," Valenti broke in, "is why did he go to all that trouble three years ago, and then just kill her this time?"

"Good question. Whatever it was he made her forget the first time, she must have found it again. It would have been too obvious to keep smacking her in the head and then wiping her memory each time she rediscovered whatever it was he was trying to keep secret, so he needed a more permanent solution."

"Well, all I can say is it must have been pretty rare," Doris said, "since it took three years to find it again."

"Not that rare." Jimmy ran a hand through his perfect hair. "Remember, they cover one block of that ten-by-ten grid each month. So going back and forth across that grid, they'd have been about forty miles away from the first spot when he killed her."

I looked at him. "*Somebody's* been thinking too much."

"Guilty," Jimmy said. "I haven't been able to *stop* thinking about it."

Doris waved a hand. "Wait a minute. What if they search the grid in a spiral pattern? Nine blocks over, nine blocks up, nine blocks to the left, then nine more back down." She tracked the pattern through the air with her finger.

"I'll be damned," Jimmy said. "Back where they started in thirty-six months."

It seemed Doris had been doing a little thinking too. I gave her a smile. "OK. So the time-line makes sense, but what could she have found that he'd have been willing to kill her for?"

"Who the hell knows? All their data is classified and it's a little tough to interrogate someone 240,000 miles away. Even tougher to go check out the spot they were surveying."

"If you can convince them to search his stuff, find that stereotactic frame, then maybe they'll give you access to their data. Or at least let you question him."

"I don't know. They're a pretty close knit bunch up there. They'd have to be to survive what they're going through. I'm not sure they'll be willing to help us pin a murder wrap on him, even *with* that CAT scan evidence. There's no proof he's the one who put those holes in her head. Now, if we can establish some kind of motive…"

"And I suppose *his* financials were clean too? Nothing that could pass for a payoff, a murder-for-hire kind of thing?"

"Nah. The only odd thing was that he was getting a monthly direct deposit into his account from Randall Corp."

That got my attention. "Interesting. So why did Randall Corp keep him on the payroll after he started working for NASA? Do you think the space agency knew about it?"

"It turns out they did. NASA let Parsons sign an agreement with Randall giving them an exclusive to his story and any footage he shoots on the far side of the moon. In exchange, NASA gets a hefty sum for anything Randall decides to use. It's a pretty sweet deal for the agency. I suspect they encourage all their lunar geologists to sign up, because every one of them has the same deal with Randall, even Susan Baker. Or she *did,* anyway."

Doris went into the kitchen to make some more coffee. She poured some water from the cooler into the coffee maker, and came back humming that damn jingle.

"Jesus, Doris. Not again."

"Huh?"

"That damn jingle's going to be bouncing around inside my skull all day."

"Sorry. I didn't even realize."

Jimmy laughed. "You two might as well get married if you're going to fight all the time."

Doris gave him the death stare. "Not funny."

He cleared his throat. "Anyway…I'm going to head back to the station, see if I can push my way up through the chain of command to put some pressure on NASA and get them to search Parsons' stuff."

"Let me know…"

"You'll be the first," Jimmy called back just before he closed the door behind him.

Doris watched the door shut, then turned to me. "I'm going to go take a shower."

"Wait," I said. "There's something I need to say."

She stood, waiting.

"Please." I motioned to the empty seat across from me, and she sat. "I've been thinking a lot about this. Maybe we should give it another try. It just feels right being with you."

She fidgeted in her chair. "Those feelings aren't real, Roger. You haven't been your usual all-controlling self the past couple days. You actually needed somebody else's help for a change, and I happen to have been the somebody. People get emotionally attached to their nurses all the time."

"Come on. It's more than that. You feel it too."

"Either way, this is not the time."

"What *better* time? We've got nowhere else to go, nothing else to do."

"That's exactly the point. We shouldn't be making decisions like this at a time when our emotions are wrapped around what we've just gone through, what we're *still* going through."

The door bell sounded—someone down at the lobby. I hesitated, then called out, "Open intercom."

The blue light around the intercom speaker next to the door began to glow. "Who's there?"

"It's Danny, Dr. Bennett. That time of the week."

I'd forgotten. The Celestial Springs guy always came by with my water delivery on Saturday morning, but with all the commotion, it wasn't exactly at the top of my list of things to remember.

"Right. Hang on a second." I buzzed him in.

Doris stood. "We'll talk about this later."

I nodded and watched her walk down the hallway, then cracked the door open to tell Valenti's guy Danny would be coming up.

As I poured myself another cup of coffee, the phone rang. It was Jimmy.

"Miss me already?"

"Actually, I've seen so much of you the past two days I don't know how much more I can take."

"What's up?"

"Well, as I started to drive away, I found myself humming that jingle Doris had been humming."

"Yeah. Annoying, isn't it?"

"Let's just say it serves its purpose, but it didn't hit me what it was until I heard the commercial on my car radio. "Nectar to the moon and the brightest stars," Jimmy started to sing.

"No. Not you too."

"It got me thinking," Jimmy said. "Remember all that hullaballoo about ten, fifteen years ago when Bucky Dumont bought the space station, the old one NASA was going to junk."

"Yeah. Nice to have so much money you can start your own space program, huh?" Dumont had made a killing selling spring water; he'd finagled exclusive contracts to supply every major Hollywood studio and their 'brightest stars.' His company, Celestial Springs, became a household name with their 'Nectar to the Stars' campaign. Hell, even I had signed up to get the stuff.

"You remember what he was going to do with it? What he *did* do with it?"

"Sure." Being the Science Fiction aficionado I was, I remembered it well. Dumont's fortune had given him the means to fuel his passion for life-on-the-edge adventure. He'd bought the old space station and turned it into a hotel for the stars, the rich and famous who could afford a week up in space. Probably not much fun when you think about it, but definitely a means to bragging rights. "That hotel—I hear it's booked-up for years."

"Yeah, in fact, I was flipping through the channels the other night and *Lifestyles of the Filthy Rich* was doing a show about Dumont. He's been shuttling these prima donnas up there every week for the past decade, and every trip up, he takes hundreds of gallons of water. He's been stashing it in these huge storage tanks. Apparently, he'd originally planned to turn it into some kind of cosmic service station; figured everyone would eventually be flying around the solar system, and when they did, he'd be there to provide water and supplies—for the right price, of course. *Last service station for one hundred light-years* kind of thing."

I couldn't suppress the snicker. "Yeah. That worked out. And this guy's richer than the Pope?"

There was a soft knocking sound from the entryway. I looked up and saw Danny peering at me from around the partially opened door and motioned him in without breaking from my conversation.

Jimmy cleared his throat. "Well, for a while there, he sure wasn't. Between starting up his own shuttle and refitting the space station, he burned through money faster than Congress; only he couldn't print more when he needed it. By the time he was ready to open his hotel, the well had run dry. And this is where it gets interesting: he went groveling to an old college buddy for a bailout. I'll give you three guesses who it was."

"Not a clue."

"Richard Randall."

"No shit. Randall? Guess it makes sense, with his interest in the moon base and all."

"Randall bailed out Dumont for a fifty-one percent stake in the company, but kept his partnership silent and let Dumont stay on as the talking head for Celestial. With Randall's money, they were able to run the hotel at a break-even, and add a whole new marketing twist to their 'Nectar to the Stars' campaign."

"Hardly seems like that would bring in enough cash to be worth a multibillion dollar investment."

"Remember who you're talking about here," Jimmy said. Randall's name was synonymous with investment savvy. "See, when NASA built MBA, they needed a way to get water there, and Randall was sitting on this huge stash of it stored in the tanks outside his orbiting hotel. It was cheaper for NASA to buy what they needed from him rather than hauling it up from Earth. It was still expensive—*very* expensive—but cheaper than the alternative. Randall made a fortune. Still does, apparently, since they resupply MBA from his tanks."

"And the rich keep getting richer."

"Right. But what would happen if one of those geologists up on MBA found a subterranean supply of water on the far side?"

"Randall would stand to lose a bundle." I felt Doris's nails digging into the flesh of my upper arm, looked up and followed her gaze to where Danny was sitting on a kitchen stool with an energy-pulse pistol aimed at us. "Oh, *shit*."

"The moment of realization, huh?" Jimmy sounded proud of himself for leading me to the obvious conclusion that had come to me much too late.

"No. The moment of, 'Oh shit, I just buzzed the Celestial Springs guy into my apartment.'"

Danny waved the gun. "Hang it up, doc."

I heard Jimmy's voice on the other end, but whatever he was saying didn't register. I tapped off the phone and put it down.

I pried Doris's hand off my arm, and held it in my own.

"You don't have to do this, Danny."

He used his left hand to help steady the gun. "They got my wife, doc."

"Jesus. Who *are* these guys?" It was Tarin Green all over again.

"I don't know, man." His voice trembled. "I got home yesterday and she was gone. They called and showed her to me—tied to a chair and gagged." He choked back the tears. "Told me where to find the gun they'd left at my house. It's either *you*," he waved the gun at me, "or her, they said."

My grip on Doris's hand slipped from cold sweat, but my demeanor remained constant and my voice didn't waver, a trick that comes from a lifetime of dealing with high pressure situations. "You don't have to kill anybody, Danny."

"I'm not going to lose her," he snapped.

I held up a hand to try and calm him. "There's another way, Danny. Just listen to me. Those guys who put you up to this, they think I have some evidence the cops need to nail them. But the cops already have it. Killing me won't accomplish a thing."

"It'll save my wife."

"You don't know that. What if she's seen them? You think they'll just let her go?"

His jaw clenched as he searched his mind for a solution that wasn't there. "Then what the hell am I supposed to do?"

"First, put the gun down, Danny. I've got friends in the Police Department. They'll know what to do."

His left hand dropped away and he slumped forward, eyeing the floor in front of him, then jerked himself upright and extended the gun in our direction. "Uh-uh. No! If I let you go, she's dead. If I do what they say, at least there's a chance."

I waved my hands at him. "Think about it, Danny. If you shoot us, she's already dead. They've got no reason to let her go and risk getting caught. She'll be gone and all they have to do to get rid of you is to wait until the cops nail you for my murder, which they will. Have you noticed all the security cameras around this place?"

"His eyes darted wildly around the room. Then *what*?"

"I've got an idea, but we'll need to buy a little time. Did they give you a contact number?"

He nodded. "They told me to call when the job was done; send them a picture of your body."

"Good. Call them. Tell them I refused the water delivery today and re-scheduled for Monday. That'll give us forty-eight hours."

"She's got to stay with that animal for two more days? Are you nuts?"

"There's not much choice here, Danny. Think about it. If you don't do this, we're all as good as dead whether you shoot or not. But as long as they know *I'm* alive and *you* can still get to me, they need you, they need your *wife*. They won't hurt her."

Danny sat silently for a moment, then flipped the gun around so he was holding it by the barrel and held it out to me. "Just take this God-damn thing, would you."

Doris slumped back into the sofa and closed her eyes. I patted her on the thigh and went over to get the gun, stuffed it into my pocket.

"Better make that call out by the window in the hallway. It's not going to be very convincing if he sees you're in my apartment." I gave him a pat on the back and walked him to the door.

The hall was deserted except for Valenti's watchman, slumped lifelessly back in his chair.

"Nice touch," I said, motioning to the baseball cap that sloped down to cover half his face.

Danny paused and looked at him. "He'll be OK in an hour or two."

I leaned on the door jamb and watched as he made his way down the walnut trimmed corridor, footsteps muffled by the tightly woven mauve carpet that paved the way. "Save the conversation," I called after him. "We need a recording of his voice."

He waved an arm without turning back.

Doris came up along side of me. "Aren't you going to stay with him?"

"He's got nowhere else to go."

He turned to face us, so the glare of light from the window obscured any background that might appear on the screen at the receiving end of the call. We ducked back inside to avoid any possibility of being seen in case he moved around.

Within a few minutes, he was back, rubbing the moisture out of his eyes.

"She OK?" Doris asked softly.

"Depends what you call OK."

I gritted my teeth. "We'll get her, Danny."

The despair in his eyes spoke volumes. "You've got your forty-eight hours."

*

As I sat across from Danny and Doris, who were waiting to hear the brilliant plan I hadn't yet concocted, a ring tone sounded from the hallway and I went to have a look. It was coming from the jacket pocket of our snoozing guardian.

I fished it out without waking him. "Hello?"

"Roger? That you?" It was Jimmy's voice.

"Yup."

"Jesus. I was about to storm your apartment. What the hell happened?"

I updated him and sent him a copy of Danny's phone call.

"A scumbag like that probably has a record. Shouldn't be hard to put a name and face to that voice. Good work."

I bounced a few thoughts off him, then went back inside with a little better idea of how we might all survive this.

"You've got to get going, Danny."

"No way. Whatever's going down, I want in on it. I want to tear that bastard apart with my bare hands."

"For this to work, they'll need to think that everything is OK; that you were never in here today; that you just made your other deliveries in the building, and then continued on with your route."

He stood. "You just let me know the second you've got that sonofabitch, the second my wife's free."

"Of course."

He grabbed the dolly he'd brought the water in on and headed out the door. I turned to Doris. "Get your stuff together. We've got to get up to the roof."

"Excuse me?"

"Jimmy's sending a chopper for us."

"Great. So now I'm a Bond girl."

"You can call me James, if it makes you happy."

"It's not all in the name, *James*." She gave my sagging belly a pat, then went back to grab her bag from the bedroom.

*

We stepped out into the brisk autumn breeze that whipped across the roof-tops of the city's highest skyscrapers and erased the defining odors of the streets below. It took a minute to adjust to the brightness of the crisp blue sky, which was marred only by an occasional wisp of cloud. By the time we had, the distinctive whir of a helicopter was coming in from the south. I waved both arms and squinted through the glare reflecting off the cockpit window, trying to see who was inside.

As the chopper approached, it started to list from side to side. I couldn't tell who the pilot was, but the other face was Jimmy's; he waved frantically for me to get out of the way as the helicopter began to jerk more violently. It cleared the edge of the building and went into a spin. I grabbed Doris by the hand as it dove toward us, and pulled her into the stairwell. A few seconds later, the sound of metal grinding against cement confirmed the worst. We waited for the engine to die, then hurried out to check on Jimmy.

He'd been thrown clear of the wreck and was screaming in pain as he tried to distance himself from the imminent explosion. It only took a glance to see that the pilot was beyond help. Doris and I dragged Jimmy to the relative safety of the concrete pillar surrounding the stairwell. I tucked my jacket under his head.

He could barely keep his eyes open as he struggled to pull his clenched hand out of his pants pocket. He let out a guttural, "Here," as he thrust the Net-pad at me, then passed out.

When you call 911 with an 'Officer Down' report, it gets attention. Within 5 minutes, a medevac helicopter was there to whisk Jimmy off to the hospital. As the deafening roar gave way to the stillness of the day, I realized that I was still clutching Doris close by my side, staring into the space where the heli-copter had recently vanished.

"You can let go now," she said softly.

"Oh. Uh…sorry."

"No need to be." She touched my cheek.

The warmth of her hand calmed me for only the briefest of moments. "Oh. Shit." I pulled the Net-pad out of the pocket I'd shoved it into as I watched over Jimmy.

I powered it up, and it opened to a picture of a middle-aged man with thinning brown hair and a pock-marked complexion. "Charles Tucker," I read the caption aloud, "1433 East Rutherford Place."

Doris strained to see the picture on the tiny screen against the brightness of the day. "That the guy?"

"Got to be."

"So what now? Does Jimmy have a partner we can call?"

I shook my head. "How do we know who did *that*?" I motioned to the smoldering wreck of the police chopper. "We do this ourselves."

"Maybe we should call Danny. We need all the help we can get."

"No. He's liable to bring more trouble than help. They've got to be following him."

"Jesus, Roger. How are you and I supposed to go after creeps like that?"

"We've got no choice. Either we get *them*, or they get *us*."

"You've got to be kidding. *Look* at us."

"Ah, come on. It'll be fun." I sounded even less convincing than I felt.

She looked me in the eye. "You're an idiot."

I shrugged.

She grabbed my hand and led me back into the building. We stopped by the apartment, where we did our best to change our appearances: a baseball cap for me, a shawl for Doris and a couple of pairs of dark sunglasses. Not likely to fool anyone really paying attention, but it was the best we could do under the circumstances. I grabbed Danny's gun, then led Doris out through the sub-basement corridors that connected three of the high-rises on my block.

We hailed a cab to take us to Rutherford Place, not the kind of place cabbies tend to frequent.

*

Charles Tucker's rap sheet had been spelled out on Jimmy's Net-pad: petty theft, car-jackings, small time drug busts, stolen-goods deals; you name it. Not the kind of guy you want to call a friend, but not the kind who masterminds a major corporate operation either. He appeared to be in unfamiliar territory with kidnapping, as far as I could tell. That was something that could work to our advantage. On the other hand, it must just make him all-the-more jittery and liable to shoot without asking questions.

The Deekin Park section of town had once been a middle class Mecca, but over the decades it had fallen into the hands of families too tired from struggling for survival to care about whether men like Tucker moved in next door.

We had the cabbie drop us off a half block from Tucker's place, a small row-house with dirty green paint peeling off aluminum siding, hiding amidst a long line of similar structures crowded along the tree-lined hill. We kept our hands in our pockets as we made our way up a cracked concrete walkway, watching, but not stopping as we passed Tucker's house. At the top of the hill, we turned right and circled back around behind the homes through a dirt alleyway.

The small yard behind Tucker's house was half driveway, half overgrown weeds where there used to be a lawn. A one-car garage sat back from the house, and a set of warped wooden steps led up to a small porch behind the back door.

We kneeled behind a row of weed-threaded azaleas, which were flourishing despite obvious neglect, and peered through a dirty window into the basement. Someone was in a chair in front of a small table, with a second person standing over him or her. It was hard to make out any details.

I waved toward the stairs, and Doris nodded. The first step let out a creak that echoed through my bones. I froze. Doris motioned for me to stop, then made her way up the stairs, keeping her weight balanced near the edge of the stairway over the underlying support beams. I lowered my weight onto my back foot, then followed her path up to the back door and dug the gun out of my pocket, trying to steady it in both hands.

"The safety," she whispered.

I tapped the OLED display on the side of the handle, but my shaky finger couldn't seem to find its way through the menu to unlock the safety. Doris took it from me, deftly tapped the screen twice, then slid the manual safety lock down and studied the readout. "Power's down to ten percent," she said.

I wasn't sure how many shots that translated into, but it couldn't be good.

Doris tapped at the menu again, pushed a button on the handle, and the battery pack slid out the bottom. She shook it vigorously for about 10 seconds, then slid it back in and gave it a smack with her palm to click it into place. She rechecked the meter, then showed it to me: 23%.

She held the gun out in my direction, handle first.

"You keep it," I said. It was obviously in better hands. "Anything else I don't know about you?"

Her eyes said, *Maybe if you'd ever taken the time to ask.*

An unlocked door would have been too much to hope for. She winced as her attempt to turn the handle failed.

"My turn." I reached into my pocket and pulled out a black plastic key fob with a silver loop dangling from one corner. "I borrowed it from Jimmy's belt before they whisked him away."

I pressed a little button on the edge of the gadget, slid it down, and a screen lit up. I studied it for a second, then tapped the activation icon. A metallic rod about the thickness of a pen-tip slipped out from the center of one side. I inserted it into the lock and tapped on the screen again. A red light started to blink at me, and a very long 10 seconds later the flashing signal mercifully turned solid green. I smiled and gave the fob a twist. The lock opened with a satisfying click.

We crouched on either side of the door and pushed it open slowly, then froze, listening to the silence. Doris led the way into the small cluttered kitchen, gun first.

We had a clear view of the door to the basement, which was partially ajar. Apart from the muffled sound of two voices—one male, one female—wafting through that opening, there was no indication of anyone else in the house, as far as we could tell. Tucker hardly seemed like the domestic type and I was pretty sure no woman had set foot in this place voluntarily in a very long time.

Doris motioned toward the coffee table in the living room across from the basement door. A tattered brown and mustard-yellow checkered sofa with matching loveseat formed an el facing us and surrounded a multicolored berber rug, so marred by overlapping stains that it was impossible to tell what it had actually looked like when it was chosen to grace the hardwood floor it now protected. An old-fashioned LED TV suspended from the wall across from the sofa featured some daytime TV talk show featuring losers that were almost as pathetic as Tucker. He must have been interrupted by a cry from his hostage in the basement; a half-eaten sandwich sat on a paper plate at the edge of the coffee table, and next to that, his cell phone.

Doris leaned in to me. "Call it."

I pulled out Jimmy's Net-pad and tapped on the number. The cell phone on the coffee table rang with the traditional default ringer used by technophobes who never bother reprogramming their phones. A few seconds later, we heard footsteps lumbering up the stairs from the basement, and my heart began to race.

I slinked back out of Doris's way, kneeling behind her. She seemed oblivious to the pressure, standing poised with her gun leveled in the direction of the door as it creaked open. There were things we needed to talk about when this was all done. The footsteps were louder now, more crisp; before the sound of the third stride echoed our way, a single pulse flashed from the handgun, followed by a loud shriek and a dull thud.

Doris watched vigilantly. She didn't move a muscle.

After several long seconds, I nudged her and gave her a *Well?* look.

She nodded and stood, put a finger to her mouth and motioned at the stairs. "Stay with him." She made her way up to the second floor, pausing slightly at each step, gun steadied in both hands.

After watching, dumbfounded, until she was halfway up, I looked at Tucker, then surveyed the room. A baseball bat was propped against the wall in one corner. I grabbed it and stood vigil, glancing up the stairs every few seconds.

Doris completed her search, then reappeared at the top of the stairwell. "Clear."

I gave her a nod, then looked back at Tucker, still out cold.

She made her way back down to me and grabbed my arm. "You OK?"

"Yeah. But how in the hell did you…"

"Later." She went down to the basement and emerged a couple minutes later with Danny's wife, who was still rubbing the sting out of the rope burns that marked her wrists.

I looked through the strands of long black hair at eyes still swollen from fighting back tears for the better part of the past 24 hours. The poor thing couldn't have been more than twenty-five or twenty-six, but the ravages of stress had taken their toll on her youthful vitality.

"You OK?"

She bit back her lip; didn't say a word.

"I'm Roger Bennett."

She muttered softly, "Jennifer," with a face so devoid of expression she could have nearly passed for a mannequin.

Doris tossed me a roll of duct tape she'd found in the basement, and then helped Jennifer over to the sofa and sat down beside her. Jennifer slowly pulled her hair back off the white floral blouse that came just shy of meeting a pair of designer jeans snuggled up around her hips, and mechanically twirled the silky tresses into a make-shift pony tail, bringing it around in front of her left shoulder.

"Thank you," she whispered, staring blankly into the space in front of her.

"Danny sent us," I said.

She turned to me and her eyes welled up with tears.

Doris took her hand and sat down beside her, and Jennifer buried her head in the shelter of the black rayon turtleneck that hugged Doris's body.

I finished the job on Tucker, wrapping the tape tightly around his ankles, and pulled his hands behind his back; left him lying on his side facing the sofa. A can of soda sat on the coffee table next to the sandwich. I picked it up and spilled what was left onto his face, then poked at him with the tip of my black leather sneaker until he started to wake.

"Who hired you, asshole?"

He forced one eyelid open and looked at me. "Who the hell are you?"

"The one asking the questions. Who do you work for?"

"Aren't you going to read me my rights?"

Jennifer stood and walked over, gave him a nasty kick in the right thigh. "Yeah. You have the right to answer our questions or the next kick'll be about eight inches higher, *dickwad*."

He winced and let out a muffled yelp, then looked up at me. "Jesus Christ. Do your job, would you?"

"What, you think we're cops?"

She drew back her foot.

"OK. OK. Just stop that crazy bitch."

"Who you calling *bitch*?" She swung her foot at the same spot, harder this time, and was answered with a satisfying shriek.

"All right, all right. Look, it was nothing personal. Just trying to make a living, you know?"

I held an arm out in front of Jennifer, kept my focus on the pock-faced man writhing on the ground. "So who was it?"

"Christ, man. He's going to kill me."

"Probably. And if you don't tell me who he is, then when I find him—and I *will* find him, sooner or later—I'll make sure he thinks you're the one who led us to him."

"If you're so damn sure you can find him, then what do you need me for?"

"I don't. But it'll make my life a whole lot easier. Just give me a name, a number…something we can find him with and I'll let you go, give you a fighting chance to get away."

"What!" Jennifer got ready to swing her foot again. "After what he did to me?"

I stopped her, took her by the hand and walked her into the kitchen. "Look. We need to get whoever it was that wanted you taken. After we're done, if Danny and you want to go after him, I doubt this dim-wit will be too hard to find."

She bit back her lip, seething, but eventually acquiesced. We went back into the room.

"So what do you say? Going to give me that name, or do I let the little lady here warm up her kicking leg?"

He motioned toward the coffee table. "The phone. It's in there. The guy's name is Bud Zielke."

"He gave you his name?"

"Not exactly, but with his number and voice print, it wasn't too hard to check him out. I just Swaggled him on the Net. Can't be too careful these days. Always good to know who you're working for in my business."

"And yet, here you are."

"Yeah, well, he was good for the money, anyway. He runs security for Randall Corp. That's about as respectable as anyone who hires a guy like me ever gets. Kidnapping's not my thing, but I figured a guy like him…"

"You figured wrong."

"No shit."

I looked at the ladies. "Come on."

"You just going to leave me here like this?" Tucker's voice had jumped an octave.

There was a knife on the kitchen counter. I grabbed it and tossed it on the floor across the room from Tucker. We'll call the cops in about an hour in case you need help getting up.

Jennifer still had venom in her eyes.

"Ready?" I asked.

"Let's get the hell out of here." She started toward the door, hesitated, then walked over to Tucker and gave him the groin kick she'd promised, answered by a shriek from his writhing body. The sealed-lips expression on her face didn't change. "*Now* I'm ready." She led the way out the front door and down the peeling painted-wood steps with a limp I hadn't noticed before.

I hurried after her into the warmth of the midday sun. "You OK to walk?"

"As long as it's away from *here*."

Doris pulled out her phone. "I'll call a cab."

"It'll be quicker to take the light-rail. We passed a stop about two blocks from here on the way in." I looked at Jennifer. "Can you make it OK?"

She nodded. "Lead the way."

As I walked, I pulled out my phone to make a call, then hesitated. "So," I said to Doris, "you never mentioned that you had such an affinity for small firearms."

"Things like that only tend to get mentioned when there are actual conversations."

Ouch. I never really stopped to consider that I knew virtually nothing about Doris, outside of work. Even when we had dated those many years ago, my conversations with her were primarily aimed at the quickest way to separate her from her panties. "So can we have one?"

She shrugged. "I was an Army brat. My dad was an ordinance expert. He taught me to respect guns from the time I was old enough to understand what they were. When I was big enough to handle them, he took me shooting twice a week. Even years after he retired, we still went to the range every Sunday afternoon."

"He still alive?"

She shook her head gently. "But I still try and go out there every week. It gives me comfort." She laughed. "Comfort in gunfire. Kind of sick, isn't it?"

"Not when you put it like that," I said. "Thanks."

"For shooting that jerk? It was my pleasure, believe me."

"No. Well, yeah, for that too. But I meant thanks for sharing that."

She smiled and tucked her hands in her pants pockets.

I dialed the number for Bud Zielke that I had copied off Tucker's phone. There were two rings before he answered. "Well," the voice said, "speak of the devil."

"Now if that ain't the pot calling the kettle black."

A wry laugh. "So tell me, what gives me the pleasure of talking to the famous Doctor Bennett?"

"Well, if you wanted to talk to me so bad, maybe you should have come yourself instead of sending a bunch of inept messenger boys."

"I don't know. It seems bad things happen to those who seek you out, doctor."

"Well if you want some advice, you've been going about this thing all wrong. Instead of trying to kill me, you should be trying to recruit me. The cops already have all the evidence they need to fry you with the records from my office, and I'm the only one who can refute it."

There was a long pause. "Hang on." A longer pause, then he came back on. "The boss wants to meet with you."

"When and where?"

"Randall Towers. One hour. The guard in the lobby will tell you where to go."

I stuck the phone in my pocket.

"Well?" Doris said.

I told her and Jennifer about the meet-up.

"Are you nuts? This guy's already tried to kill you twice and you're going to his office? *Alone?*"

"They've got no reason to kill me anymore. They got to Jimmy too late; the evidence is out there."

"Then why meet with him?"

"Because I need to make sure this all stops here, that Randall's lackeys don't try and kill all of us just to tie up loose ends. I need to give them a reason to keep us alive."

"And that reason is?"

"I'm working on it."

*

Randall Plaza is one of the most frequented spots in the city, bustling with the business crowd during the week and tourists all weekend long. It was starting to get late, but the shops and restaurants lining the lobby level were still bustling with pedestrians. We ducked into a coffee shop across the plaza from Randall Towers and took a booth by the window.

The ladies got in line to get some iced cappuccino while I studied the entrance to the building. Doris came back with her drink and a wad of brown paper napkins and slid in next to me, while Jenifer sat across the table, silent, introspective. I wanted to say something to make her feel better, to erase the pain of the ordeal, but nothing could do that, not this soon after.

I felt a tugging at my jacket and reached down. Doris had stuffed the pistol, now wadded up in napkins, back into my pocket. I pulled it out, kept it tucked between our bodies and pushed it back toward her. "No, you keep it."

"You can't go in there unarmed."

"How am I going to get that thing past security."

"It's an office building," Doris said, "not an airport."

"Maybe so, but Randall's goons still aren't going to let me anywhere near him with that. Besides, I'm more likely to get myself killed if I actually try and use it. I'll just have to rely on my keen wit, boyish charm, and devilishly handsome looks to protect me."

"Let's hope that wit's firing on all cylinders." She stuffed the gun back into her purse, then stood to let me out.

"Don't sit here too long. Keep moving, stay in the middle of the crowds and blend in. If you don't hear from me by the time it starts to get dark, go over to the Royal Arms and ask for the manager, Ed Traschel. He's an old buddy of mine. He'll check you in under a false name. I'll find you there."

As I turned to go, Doris took my hand.

I looked into her eyes. "I'll be fine."

I hoped I was right.

*

A guard perched at the concierge desk just inside the Randall Towers entrance directed me to an elevator at the far end of the lobby, around the corner from the main bank of elevators. The security eye recognized me and opened the door. "Good afternoon, Dr. Bennett," it greeted me with the pleasing drawl of a southern belle.

"Hi, there." Stupid, I know. But these damn things were always so polite, I felt myself obliged to respond.

I entered and the door closed behind me. There was no perceptible motion for five or six seconds as a dull hum crept up the wall from floor to ceiling—

the unmistakable hum of a scanner. I'd heard it in the airport screening booth a hundred times before, but never in an office building. I'd made the right choice leaving that gun behind.

A few seconds after the hum stopped, the car started its swift ascent without inquiring which floor I wanted to go to. The door opened into the private penthouse office of the world's richest man. Plush rose carpet blanketed the expansive oval-shaped anteroom. Two enormous picture windows following the curve of the wall in front of me separated us from the city outside, the view disrupted only by a solid section of wall between them, decorated with textured wallpaper and the giant eagle logo of the Randall Corporation, carved in multicolored woods. Just in front of that wall and matching the shape of the room, an oversized desk of rich dark oak overwhelmed the statuesque woman who sat behind it, not such an easy task considering that she'd obviously been hired for her physical attributes.

After briefly freezing at the enormity of it all, I exited the confines of the lift.

"Have a nice day," the elevator said.

I paused and looked back over my shoulder. "You too."

That got a smile from the elegant blond behind the desk. "Dr. Bennett?"

"Yes."

"Mr. Randall is waiting for you." She stood and walked to her right, coming out from behind the massive piece of furniture. "Follow me."

I complied, trying not to focus on legs that begged to be focused upon. She knocked on a large oak door curved to match the wall it was set into at one of the narrow ends of the oval room. I guess she was tuned into the voice coming from the other side, because I didn't hear a thing. She pushed the heavy door open with ease and motioned me in.

I instinctively patted my jacket over my shirt pocket where I'd tucked my laser-knife pen, then walked in.

Standing there, facing a man you've seen all over TV and magazines covers for much of your adult life can be a bit overwhelming. I must have looked as stupid as I felt, a look Richard Randall was obviously used to seeing. I was so star-struck that I didn't even notice the voluptuous model who guided me in here making her exit.

He walked out from behind his desk and extended a hand. "Richard Randall."

"I *thought* that might be you." I forced a smile, gave his hand a firm shake.

He laughed politely. "It's a pleasure to meet you. Bud's been telling me all about you."

I tensed as Bud Zielke stood from the cover of one of the high-back chairs facing Randall's desk, and turned to greet me. He looked very much like his picture on the Net, only a few years older and very much bigger—my image of an ex-marine, now 20 lb overweight but still with a physique that mere mortals like myself wouldn't want to mess with.

"Doctor." He walked over and started to frisk me.

Randall grabbed his arm. "Whoa, whoa. Bud, what is this? Dr. Bennett is a guest here."

"Indulge me," Zielke said. He shook off Randall's grip, then continued what he'd started. A smug grin broke across his face as he opened my windbreaker and spotted the brushed metal pen in my shirt pocket.

He pulled it out and waved it in front of Randall. "This is what he used to kill that reporter, Tarin Green."

Randall cocked his head. "A pen?"

Zielke activated the laser, set to maximum, and the beam danced along the ceiling.

Randall smiled. "Looks like my presentation pointer, Bud. What's he going to do, bore us to death with a Power Point presentation?"

Zielke took Randall's zebrawood business card holder off the desk, dumped the cards and ran the laser scalpel along the base, a few millimeters from the surface. The end Zielke was not holding fell to the floor. A clean cut.

Randall's jaw clenched.

"Sorry," I said. "Been carrying that thing since someone tried to kill me for the second time this week."

"Disgruntled patient?" Zielke's eyebrow rose with pride from his attempt at wit, and he stuffed the scalpel in his pocket.

I glared.

Randall's forehead filled with wrinkles. "Who the hell would want to kill you? You're a hero."

I didn't know how to answer that. The guy actually looked serious.

"Doesn't matter," Zielke barked. "Shouldn't have brought it in here." We stared stone-faced at each other.

Randall cleared his throat. "You're right, of course." He turned to me. "Look, Doctor, I can only imagine what you must be going through, but you shouldn't have brought that thing into my office."

"My apologies."

Randall took a deep breath. "Let's try this again, shall we?" He stared down Zielke, who took an obedient step back. "Doctor," Randall motioned toward the chair adjacent to the one Zielke had been sitting in. "Please." I took a seat while Randall made his way back to his position of power behind the desk. Zielke remained standing vigil over the proceedings.

"So," Randall said, shifting gears and lighting up his best boardroom smile. "I was so excited when Bud told me you were coming. The man who performed the first operation on the moon. Right here in my office. Incredible."

I met his enthusiasm with silence. He seemed awfully amiable for someone who was about to lose the empire he'd spent his entire life building, and in large part because of what *I* had discovered.

He waved a hand. "You'll have to forgive me. I'm such a geek. How did it feel to actually operate on somebody on the moon?"

Here we go again.

"No different than if she was across the room from me in the robotic surgery suite at the hospital."

"Oh, but it *was* different, Doctor. Very different. Man, I'd love to see that surgical robot in action."

"I, uh…" I looked over at Zielke, who would have been a hell of a poker player, then back to Randall again.

"For God's sake, Bud," Randall said, "sit down, would you? You're even making *me* nervous." He let out a restrained chuckle.

Zielke kept his gaze on me, didn't move for a long moment, then sat.

"Good, then. Go on, Doctor. You were saying?"

"I'm sure we can arrange it."

"Great." He rubbed his eyes with an thumb and forefinger, then shook his head as if he was having trouble focusing.

"Late night?"

"Nah. Just too much wine at lunch, I guess. The revolving restaurant here is awesome, you should give it a try. Just flew a new chef over from Paris and he was worth every penny it took to steal him away.

I studied his face. "Mr. Randall, you *do* know why I'm here."

Zielke bolted up out of his chair. "Oh, for Christ's sake."

I looked up to see his gun pointed at me.

"Shit," I yelped. "Are you nuts? I'm your only chance here."

Zielke snarled. "I know." He turned toward Randall, who'd started to stand, and fired a silent pulse. Randall collapsed instantly, smacking his head against the desk with a sickening thud.

I was still frozen in my seat when Zielke spun and turned the gun back on me. "Enough of this God damn charade."

"Look, look," I waved frantically with both arms, "I can get you out of this mess."

"That's the only reason you're not dead yet, doc."

"Then why kill Randall?"

"Oh, he's not dead. Not yet."

I studied his face as it broke into a grin. "Somebody's got to take the fall for this and it ain't going to be me."

"So *he* ordered you to do all this?" I motioned toward Randall.

"Nobody *orders* me to do anything. Hell, that self-righteous asshole was so squeaky clean, he never would have done what *had* to be done to make that buy-out of the space station pay off. It was just a new toy to him."

"Why the hell would you care? Not really your end of the business, is it?"

"When that sonofabitch bought Celestial, I put every penny I had into it, figured if the great Richard Randall was buying it, it was a sure thing. I wasn't going to let anybody bring it down. Not him," he motioned toward Randall, "not you."

"So *you* were behind Baker's murder." As the words escaped my lips, I realized that wasn't the brightest thing to say under the circumstances.

"Hell, no. That jackass, Parsons did that. When he wiped her memory three years ago, *that* was my idea. After that, I told everyone to back off. It would have been too obvious if things kept happening up there. We'd already made a bundle."

"So why did Parsons kill her?"

Zielke started stretching his left hand and clenching it, rhythmically. I tried not to notice. "Why else. He got greedy."

"But we checked his financials," I said. "He doesn't own any stock."

Zielke grinned. "That you could find. Guess he wasn't a total idiot."

"And you couldn't turn him in because it would have implicated you."

"Nah, far as he knew, everything was coming directly from Randall. But once an investigation gets going, who the hell knows?"

"Which is why you sent Green over to wipe my files."

"Putz." His eyebrow lifted with the inflection. "Only stirred things up. Killing you wasn't high on my agenda, not unless you figured out what happened."

I fixed on his eyes, gritted my teeth.

"Don't sweat it, doc. I didn't bring you here to kill you, not unless you force my hand." As he spoke, he rubbed at the back of his neck, craning awkwardly. "See, the evidence trail may all point to Randall, but Randall will point to *me*, eventually. The only way I stay out of the line of fire is if he's dead. But if they find me in here alone with his dead body, how much chance do you think I'll have of talking my way out of that one? With you here, I've got options."

"Options?"

"Two of them. Your choice, doc. First one, I *do* kill you, then put the gun in your hands and fire the fatal shot at Randall from where you're sitting. I'll get a friggin' medal for taking out the man who murdered the great Richard Randall."

"And they're going to believe that *I* managed to wrestle a gun away from *you?*"

He laughed. "Hell, no. You looked at yourself in the mirror lately?" He pulled a second gun out of his pocket. "The kill shot'll come from this one. Your gun, your fingerprints." He stuffed it back into his pocket.

"*My* gun? I don't own a gun."

"Sure you do, doc. You bought it a couple weeks ago. Check your credit statement."

This guy was smarter than he looked.

"So what's option number two?"

"I *don't* kill you. I put the gun in Randall's hands and shoot him point blank. You back me up when I tell the cops he committed suicide after we figured out he was behind Baker's murder."

"And why would you do that?"

His eyes started to glaze over and he quickly shook it off. "Being alone in here with two dead bodies is going to lead to a crapload of questions. If it's just him, and I've got you to back me up, it'll go a whole lot easier."

An uncharacteristic wobble disrupted his staid military stance and his smug demeanor slowly morphed into the intense glare of realization. "What the hell did you do to me?"

It was my turn to grin. "I gave you another option."

He leaned against the desk, fought to steady the gun with a two hand grip.

"You get to live if you hand me that pistol and let me call 911." I held up my right hand, carefully peeled a band aide off my index finder. "Didn't have anything as sophisticated as that delivery system you gave Green to kill *me* with, but hey, you learn to improvise. I rubbed some Xetaphol on this before I came up here, something we use to put patients under for surgery. It's pretty harmless when an anesthesiologist is around to make sure you don't stop breathing." I glanced around the room. "Don't suppose you have one of those around here, do you?"

Zielke's eyes narrowed to two slits. "You son of a ..."

"Save your breath. You're going to need it."

"But I never touched your damn hand."

"Nah, I had a feeling you might not be too sociable, so I also smeared some of it on that laser scalpel you pulled out of my pocket; figured I'd only get myself killed if I tried to use the damn thing anyway." I motioned towards Randall. "You didn't need to shoot *him*, though. With that firm grip he gave me...he wouldn't have lasted much longer."

Zielke staggered, hesitated.

"And if you're thinking of shooting me just for spite, remember, I'm the only one who can keep you alive until the paramedics get here."

He snarled, dropped the gun on the desk and threw himself into a chair.

I stood and went over to push the gun out of his reach. His arm shot out and grabbed me by the wrist. Even half-asleep, the bastard was strong as hell.

"This ain't over, doc."

I took the gun by the barrel and with a single motion, brought it down sharply on his forearm. His weakened grip gave way with a satisfying crack. "Yeah. It is."

As Zielke lingered between consciousness and sleep, I hurried over to try and wake Randall. The Xetaphol dose I'd given each of them would make them pretty groggy, but that's about it. When I'd hatched this plan, I knew I wouldn't need to actually kill them, just make sure they *believed* they were going to die. I always prefer not killing people; it kind of goes against the Hippocratic Oath.

Waking Randall wasn't easy. Between the drug I'd slipped him and the stun shot he'd taken from Zielke, he was going to be pretty wobbly for a while, but he'd be OK.

I tapped the intercom on Randall's desk and yelled for security as I tried to rouse him. Just as he started to moan, three uniformed guards burst through the door and tackled me; made sense since I was standing over their two semiconscious bosses.

Fortunately, Randall came-to a few seconds later and remembered who'd shot him.

*

Starlight danced over the city, reflecting off the Merriweather River as it meandered through neighborhoods rendered indistinguishable from the height of the revolving restaurant atop Randall Towers. But as inspiring as the view was, I couldn't keep my eyes off Doris.

I barely recognized the stunning woman sitting across from me as the same one who ran my office every day. Some of it was no doubt the deftly applied make-up and simple black gown that accentuated her finer features, but most of it was the gleam in her eyes and sultry tone of a voice that had never made its way into the hospital. I'd seen her outside of work before, even in recent years, but this was more than work versus leisure. She radiated a warmth I'd never felt before, not from anyone.

The glistening city lights only enhanced the romantic atmosphere permeating the chic restaurant. I caught Doris stealing a glimpse of the gilded dome of the Gates building as it came into view. "Isn't this view amazing."

"That it is. I must admit, I didn't fully appreciate it when I was standing in Randall's office."

"Don't even joke about that." She glanced up at the ceiling that separated us from Randall's penthouse suite. "It still gives me chills every time I think about you being stuck in that room with Zielke. Your whole plan hinged on that lunatic believing that you—a *doctor*, of all people—were willing to commit murder. If he didn't fall for your bluff, he could have just killed you and pinned Randall's murder on you. How could you take a chance like that?"

"Who says I was bluffing? You don't know how tempted I was to finish the job, once I had that gun in my hand. The son of a bitch did try to kill me— *twice*."

"Oh, come on, Roger. You're no murderer."

"Hey, I killed Green, didn't' I?"

"That was different; it was self-defense."

I still had the image of Green's blood-soaked body etched into my brain. "Not so different."

"And you're OK with that?"

I took a deep breath and stared blindly down at the table setting in front of me. "It hasn't stopped haunting me; not for a minute."

She reached out and put her hand over mine.

As I looked up, we were interrupted by a waiter balancing two crystal Champagne glasses and a bottle of Dom Perignon on an engraved silver tray.

"From the gentleman," he said motioning across the room.

I looked over and saw the smiling face of Richard Randall acknowledge me with a subtle nod. I returned the gesture as the Champagne was poured, then lifted a glass toward Doris.

As we gently tapped glasses, the regal sound of fine crystal sent a scintillating chill down my spine, a harbinger of unknown passions.

It was a toast 27 years in the making, but one which could have never been made until that very moment. One that needed no words.

5

Hiding from Nobel

How would it feel to know you've made the greatest medical breakthrough in history, but that you can't tell anyone about it?

The setting of this story is a summer camp that was an important part of my childhood, but the understandably mad scientist is purely fictional. It would be really cool to meet him, though.

The years rolled by in my mind as I pulled through the gates of Hidden Meadows, an upscale development in northern Maryland. It hardly felt like 25 years had passed, yet what had happened that day seemed like an eternity ago. I drove through the now unfamiliar landscape, winding my way toward the crest of Girls Hill. At least that's what we called it back then. It was there under the shade of the uppermost tree in a row of pines that lined the edge of the hill that we had made our pact.

*

Benny Solomon swiped at the beads of sweat dripping off his short-cropped wavy black hair. "Hard to believe another summer's gone," he said, extending an index finger to readjust the horn-rimmed glasses that had slid down the bridge of his nose.

"Ah, c'mon, Solly." Zeke scratched at the fine hairs of a nascent adolescent beard that hadn't changed all summer. "We got two more whole weeks. Quit your whining."

"At least you guys get to start high school," Jeffrey said. "I got another whole stinkin' year of *junior* high." He aimed a puff of air at the strand of straight blond hair that always seemed to be dangling in front of his eyes.

"Yeah," I huffed through the best sigh I could muster up. "Hard to believe this is it. Next year we're gonna be too old to be campers. At least most of us," I grinned at Jeffrey.

For the past five summers we had been coming to Camp Ramblewood. I was just 10 years old when I started. That was Solly and Zeke's first year too.

B. Aiken, *Small Doses of the Future,* Science and Fiction,
DOI 10.1007/978-3-319-04253-4_5, © Springer International Publishing Switzerland 2014

Everyone else in Bunk Nine knew each other from prior years, so the three of us had a common bond. We became pretty tight that summer, but when camp ended we lost touch. There was no Internet back then, at least not for kids, long distance phone calls were too expensive and boys are too lazy to write letters. But when summer rolled around again each year, it was like we'd never been apart.

By the third summer, we had taken Jeffrey under our wings. We were twelve and girls were starting to look pretty good to us. He may have been a year younger, but his longish blond hair and blue eyes were like dangling bait at the Saturday night socials.

We grew closer each year and by the summer of 1985 we were inseparable.

"Let's make a pact," I said. "Twenty-five years from today, wherever we are, whatever we're doing, we meet right back here on this very spot."

"Let me check my calendar," Solly said.

Zeke shot him a dirty look.

"That's like *forever*, guys." Jeffrey was wide-eyed at the proposition. "Very cool." He put his hand out and made a fist, thumb side up.

We each followed suit, stacking our fists up in a column.

"August first, 2010, at high noon," I said.

We broke the column and reassembled our balled-up hands knuckles to knuckles.

"At high noon," we all repeated in unison, and then tapped our fists together twice before breaking ranks.

We were all so innocent then.

*

Glancing in the rear-view mirror, I wondered if they'd recognize me. I winced at the budding crow's feet around my eyes and the streaks of gray beginning to cut through my thick black hair.

"Hell, they probably won't even show up," I said to my reflection. It had been my idea, this pact. A boyhood fantasy, a moment of adolescent bonding surely forgotten by three grown men.

As I approached the spot where history was to be repeated, my heart sank. What was once nature's paradise was now suburbia. It took a few minutes to get my bearings, but the lay of the land began to bring back memories. I pulled to the side of the road and turned off the engine. Somehow it didn't seem right to drive over what was once the grassy crest of Girls Hill. About fifty yards up and to the left was the spot where we'd made our pact 25 years earlier. The pine tree that shaded us that day was long gone, as was the whole row of trees leading down the hill to the pool and L-shaped dining hall framing a part of it. What had once been a soft bed of pine needles nestled under filtered shade and sunlight was now an asphalt road.

I locked my car and walked past the meeting spot along the ridge of the hill. Cabins that had once housed the female campers had been replaced by a row of two story red brick homes, which continued around the corner and followed the contour of the gentle slope down to its base, once the center of camp activity.

The pristine land that had made up Girls Hill and Boys Hill with the dining hall in the valley between them had been preserved as a neighborhood park. Pausing by the roadside, I closed my eyes and listened to the chirping of the birds; for a brief moment I was back in 1985.

To my left, the row of pines stood serenely against time, leading down to the dining hall and then back up to the cabins on the crest of opposite hill. Those wooden shacks where I'd spent much of my youth were lined up like dominos, housing 5 year olds in the left-most building and progressing along the ridge, ending with the hormonally charged teens of Bunk 14 at the edge of the woods to the right. The dense green forest making up those woods formed a boundary that stretched down toward the pool and then back up again to the girls' cabins, which lined the land behind where I now stood, enclosing the rectangle of camp life as they connected the woods to the pine trees on this side of heaven.

And then, as they inevitably do, the memories led to *that* moment, and I winced in pain.

*

"C'mon squirt," Zeke said to Jeffrey. "You chicken?"

Jeffrey grimaced. "I ain't scared of nothing'," he snapped.

"Then come with us tonight."

We'd been planning this raid all summer. It was tradition. The senior boys' bunk would pull a night raid, sneaking up through the woods to Girls Hill long after the counselors had fallen asleep. Armed with shaving cream and toilet paper, they would decorate the cabin of the senior girls' bunk, then steal back to Boys Hill under cover of night.

"My counselor will kill me. He told us there would be hell to pay if you guys pull the raid this year."

"So, what, are you gonna turn us in?" I said.

"Hell, no."

"Then come with us. You're almost a senior camper now, anyhow."

"I don't know…"

We all knew at that moment that Jeffrey was going with us.

*

I surveyed the top of Ramblewood Lane, the street that they had paved over Girls Hill. There was no bench, no pine trees, but a young oak at the top of the hill provided a bit of shade. I could swear they had planted it in the very spot that we had made our pact. I plopped down onto the ground and leaned back against it, then glanced around hoping to catch a glimpse of a familiar face. A group of kids rode by on their bikes and looked back at me dubiously. I couldn't blame them. A 39 year old stranger sitting in the grass at a lonely suburban intersection was a strange sight. I expected to see the police shortly.

I closed my eyes and the memories washed over me once again.

*

"You ready?" Solly tapped me on the shoulder. He was our alarm clock. Solly had an uncanny ability to program his body to awaken at any pre-ordained time.

"Hmm?" I mumbled. "Is it one already?"

"Yeah. Get a move on," he whispered. "You get Jeffrey and I'll get Zeke. We'll meet by the edge of the woods."

I nodded and dragged myself out of bed. Ten minutes later, Jeffrey and I were walking behind the cabins towards our traditional meeting spot behind Bunk 14, near the edge of the woods that would provide cover for our clandestine mission. Zeke and Solly were waiting.

"What took you guys so long?" Zeke snapped.

"Hey, we're here, aren't we?"

It was dark, but there was a quarter moon that night, and I could see that everybody had their gear. We each carried a small satchel of supplies that we had readied the day before—shaving cream, a roll of toilet paper and a flashlight, which we would use only in case of emergency.

It was just the four of us. The rest of our bunkmates had decided not to chance the wrath of the camp owners who had issued the edict banning this summer's senior raid. Solly, Zeke and I had pretended to agree with them, but there was no way we were going to pass up an opportunity that we'd waited five long summers for.

"Lets go, then."

Zeke led the way down the hill, sticking close by the tree-line, concealing himself in the moon shadows. Solly and I were right behind. Jeffrey's legs were shorter, and much to his chagrin, his timidity got the better of him in the darkness of the night. He lagged a dozen yards behind, urging himself on and trying to keep up.

I glanced back a few times. "Wait up," I called ahead.

Zeke looked over his shoulder and snickered. "If the squirt can't keep up, that's his problem. It'll make a man of him." He increased his pace.

Solly and I looked at each other and shrugged. Jeffrey wasn't that far back. It wasn't like he was going to get lost or anything, and we all wanted to get the job done and get back as quickly as possible. We forged on.

*

"Catching a few z's?"

Much to my surprise, I recognized the voice immediately. Squinting up from my spot under the oak tree, I struggled to focus on the figure silhouetted against sunlight.

"Good to see you, Solly. I can't believe you came."

"We swore on it, man."

"Yeah, but that was a long time ago."

"Many moons." He plopped down next to me.

"I was beginning to think no one else would show."

Solly looked at his watch. "Right on time."

I should have known. Solly was always punctual, even as a teenage boy. "Guess I was early."

"That you were. How the hell have you been, man?"

"Life's been good to me. Gorgeous wife, two kids and a dog. The all-American dream."

"Way to go," Solly said with a smile.

"How about you?" I asked.

"Well, I was pretty messed up for a while."

"You?"

"Yeah, well, you know. I took it pretty hard. My parents did too. After I got expelled from camp, they were humiliated. That kind of stuff doesn't happen to Solomons. They sent me away to a boarding school. I ran away after 2 years of that crap."

"Geez," I muttered.

"Yeah, well. I grew out of it eventually. I went to BU and took over my old man's business. I'm all respectable now. I did the family thing too—wife, two kids; no dog, though. I'm allergic."

I laughed.

"What?" Solly feigned offense.

"Sorry," I giggled. "It's just…well, it's not too surprising, you know?"

Benny Solomon shook his head. He knew.

*

Solly couldn't stop sneezing as we tried to sneak down along the forest line.

"Geez, keep it down, would you?" Zeke snapped.

"I can't help it," Solly said. It's the honeysuckle."

The native plant was plentiful along the forest's edge.

Solly's effort to suppress nature's curse was defeated by a trumpeting blast of moisture that made his head ricochet.

Zeke spun around and gave him a dirty look.

"Look," I said, "he's allergic. He *can't* stop it. The sooner we get past the honeysuckle, the sooner he'll stop."

Zeke grunted and turned back toward Girls Hill, quickening his pace. Solly, trying desperately not to sneeze, followed close behind with me at his side. Jeffrey was falling further back, just within eyesight in the dimness of the night.

*

The shade of the oak felt good in the heat of a Maryland summer's day.

"Think Zeke will show?" I asked Solly.

"What, you didn't hear?"

"Hear what?"

"Zeke died, man. Motorcycle wreck, back in ninety-four."

My mouth opened, but nothing came out. It's not like we were all that close; hell, I hadn't seen him since we were kids, but still, it felt like a part of me had been chopped off.

"I can't believe you didn't hear about it. There was a ton of coverage. Flew his bike off an overpass into oncoming traffic on 495. Real gory accident, the kind of stuff the press just love."

"Must not have made it to the Baltimore papers." Jeffrey, Zeke and Solly lived in DC back then. I was from Baltimore. "They say anything about Zeke? What he was up to all those years?

"Nah, not much. Just that he was in and out of jail all the time. It probably was a little harder on *him* after we got kicked out of camp; no family to lean on." Zeke was never too shy when it came to talking about how his foster parents treated him.

"Ah, come on. Nothing bothered Zeke. He never even *liked* Jeffrey that much."

"Guess he had us fooled."

We sat quietly, and I was sure Solly was thinking about the same thing as I was: the tough kid who pushed us to the edge of trouble but never let us fall in. I had always admired the fact that nothing rattled him. I guess we all have our breaking points.

"At least it was quick," Solly said.

"Yeah."

"And at least he didn't have to live with the guilt anymore."

I looked at the stress lines on my old friend's face. "That part *would* be nice, wouldn't it?"

Solly nodded.

"The nightmares don't come as often, but they still come."

"Yeah, I know what you mean."

"We never should have let him get so far behind…"

*

Solly had given up the futile effort of trying to suppress his inevitable reaction to the omnipresent honeysuckle and let loose with a nice long, loud one. We were half way up Girls Hill by that time, well within earshot of the cabins.

Zeke spun on his heels, right into Solly's face.

"Shut that snout of yours, Solomon, or I'll shut it for you!" His fist was clenched.

"Come on, Zeke," I pleaded. "It's not like he's doing it on purpose."

"It don't matter *why* he's doing it," Zeke said. Hell, every counselor on the hill's probably up by now."

We all glanced up toward the girls' cabins, expecting to see flashlights pointed our way.

And that's when it happened.

A hideous roar. Surreal. One vicious growl, and then a yelp of terror from Jeffrey. We strained to see into the night. Solly was the first to get his flashlight out and just as Zeke was about to yell at him for turning it on, we saw something dart into the underbrush at the edge of the woods. The beam of light followed its path to two red eyes trained on us from between the branches, and triggered a malicious snarl that seemed to rumble back along the trail of illumination.

"Wolf!" Solly rasped.

We stood motionless. We'd heard rumors of wolves roaming the woods, but always figured it was the counselors' way of keeping us out of the forest. Nobody had ever actually seen one.

Leaves rustled from behind the wolf and a sound…like a swarm of crickets…rose from the area. I was too scared to run. We all were.

Then, without even acknowledging our presence, a man whose considerable height was amplified to near cartoonish proportions by his wiry build strode calmly out from the bushes. His gaunt face was devoid of features in the darkness of the night, with a long gray beard the only hair decorating his odd visage. A brown plaid trenchcoat draped loosely on his frame hid any other details we could have used to describe him later. As he walked toward Jeffrey's supine body, I felt as if I were in a dream, watching from a precarious position that could have been inches or lightyears away, but unmistakably close enough to be pulled into that nightmare in an instant. Frozen, helpless to retreat, I was resigned to the feeling that no distance would be a safe one.

When he reached the spot where Jeffrey was sprawled out, he stopped and the wolf edged out of the forest to sit obediently at his heels. The man knelt over Jeffrey and raised his left hand, scanning it over the motionless head and torso, which glowed a faint tint of blue at the effort. The light went out for an instant, and then the pale blue haze poured back over my friends face. In the stillness, I could not even feel myself breathe, and then Jeffrey started to stir. At the very same moment, the man stood his six and a half foot frame upright, turned back toward the woods at an angle that continued to shield his face from us, and disappeared calmly into the woods with the wolf loping along behind.

Zeke shined his light in their direction. "Gone. Who the hell was that guy?"

"Who cares," Solly said, training his light on Jeffrey. He and I scurried over to check on Jeffrey, while Zeke continued surveying the trail, or lack of it, where the stranger and the wolf had vanished.

The rock under Jeffrey's head was covered with blood, but the stickiness had stopped oozing from him. He winced as he tried to raise himself up, and we helped him sit.

He began to mutter something, but passed out before we could make out what he was saying.

As we settled him back down on the soft grass, we were blinded by the headlights from a camp security jeep. It stopped a dozen yards away and by the time my retinas had recovered enough to allow me to focus, we were surrounded by a mob of counselors, campers and two gray-haired security guards with glasses like Mountain Dew bottles, who looked more frightened than most of the children.

It was all a blur, what happened after that. One of the counselors motioned for us to get away. We couldn't bring ourselves to move, but as we were dragged from the scene, I could see her bending over my friend's limp body. A helicopter arrived about 15 minutes later and whisked him off to a hospital in Wilmington.

Nobody believed our story. Jeffrey didn't have a mark on him besides the gash in the back of his head, and there was no trail, no sign of the wolf or the skinny giant. They figured we'd coerced Jeffrey into coming with us, and then didn't look out for him like we should have. That we made him run to keep up, and then when he tripped and whacked his head on a rock, we concocted that crazy story about a wolf and a giant. They were right except for the last part, of course, and we had no proof.

The next day, the three of us were picked up from camp by our angry and embarrassed parents.

*

Solly and I sat under the oak in silence, thinking about the day that had changed our lives.

"Did you ever find out?"

"Nope," Solly said.

"Me neither. My folks wouldn't let me have anything to do with camp anymore. Not the place or the people. I figured Jeffrey must have died or they would have told me, not let me suffer so much. But I think *they* never really knew either."

"Did you ask them?"

"Nah. It was too painful—for all of us. We never mentioned it again. I check the Internet every once in a while, you know, to try and find out what happened. But I always come up dry."

"Maybe just as well," Solly said. "Sometimes the past is better left in the past."

We sat back against the tree again in silence. No one else was left to come to this reunion today.

"But sometimes it's *better* to know." A lilting feminine voice wafted in from the other side of the oak tree.

Solly and I both spun around.

"Mindy?" I couldn't believe my eyes.

"In the flesh," she smiled serenely.

Solly and I both stood. Mindy was my first girlfriend. A summertime fling for a 12 year old is not a serious thing…except to that particular 12 year old.

"But…"

"What am I doing here?" she finished my question.

Solly and I both nodded.

"Jeffrey sent me."

My jaw dropped.

She smiled, but with a tear in her eye. "He wanted so much to tell you. He never blamed you for what happened."

"Where the hell has he been all these years? I Googled him, tried Facebook, even tried to track down his family…nothing. It was like he'd dropped off the face of the Earth."

"In London," she said matter-of-factly. "With me."

"You?"

She took a deep breath. "After the accident at camp, it took Jeff a while to recover. Physically he was OK, but he had panic attacks, Post-traumatic Stress Syndrome, they called it. He struggled in school and became alienated from his friends, so his parents moved to London for a fresh start. Eventually, he found his way, became a psychologist.

"About 10 years ago, I was vacationing in London with some friends and spotted him sitting at a pub in SOHO. Even after all that time his face had hardly changed. We started talking and one thing led to another; we were married a year later."

"We had a quiet life there. Then one day, about a year ago, it all changed. A new patient walked into Jeffrey's office and, well…it'll be easier if I just show you. Jeff keeps video records of all his encounters."

Mindy pulled out her iphone, started the video clip and handed it to me. Solly sidled up to get a better angle. We could only see Jeffrey's back, but the view of the patient was plain as day, even on that tiny screen.

The wiry man who walked into that room was so tall the camera angle cut off the view of the top of his head. He sauntered up to the desk with a deliberateness that conveyed a complete disregard for the constraints of time, and sat. His long gray beard was the only hair visible, and a loose-fitting brown trenchcoat hid the details of his frame.

"Shit," I muttered. Turning to Solly, I got the confirmation I dreaded.

Even on the iphone's tiny screen, the image immediately reactivated the feebly suppressed memory of the Ramblewood hermit who had revived Jeffrey that night. His steely gray eyes were mesmerizing, and he looked considerably younger than I had imagined, despite deeply set cheeks and pale, nearly albino skin tone. After all these years, I finally had a face to put to that gaunt profile.

The sound of Jeffrey's voice coming from the iphone drew me into the conversation. "Mr. Zile?" He extended a hand. "I'm Dr. Blondell."

The man shook his hand and nodded. Jeffrey motioned for him to sit and they each settled in on opposite sides of the bean-shaped oak desk.

"My name," the man started in an authoritative, deeply timbered voice that was contrary to any I would have imagined coming out of him, and I had imagined a great deal about this man over the years, "is not Zile, but it's best for both of us if you do not know my true identity."

Jeffrey's head tilted. "Look, Mr…whatever your name is, if you're not going to be honest with me, I can't help you. Whatever you tell me in this room is confidential."

"But I don't need your help, *you* need *mine*."

Jeffrey rocked back in his high-backed leather chair. "You're here to help *me?*"

"I am."

"OK," said Jeffrey, "I'm listening."

The man studied Jeffrey's face. "You don't recognize me, do you?" After a brief pause, he answered his own question. "No, I suppose you wouldn't. You were dead the last time we met."

"Dead," Jeffrey repeated flatly. "I was dead?"

The man nodded. "Only briefly."

"Don't recall ever being dead," Jeffrey said, with a tinge of amusement coloring his voice.

"1985. Northern Maryland. Camp Ramblewood."

Jeffrey leaned in and rested his hands on the desk, fingers intertwined. "I don't appreciate your dredging up my past, Mr…"

"Zile will do."

"Fine. Mr. *Zile*. I think it's time for you to go."

Jeffrey stood, but the man in the chair did not budge.

"I was there, Jeffrey. I was the one who revived you."

Jeffrey sat back down. Even if he *had* heard our version of what had happened that night, it would have been relayed to him in a tone tainted with the doubt of those who had pegged us as liars. We'd been told at the time that Jeffrey didn't remember any part of what had happened, and probably never would. He had no reason to believe our bizarre story. Until now.

"My name is not important, but I'll need to tell you a little about myself for you to understand *how* I saved you, and why it's important only *now* that you understand."

"When I was a young man, I studied theoretical physics at Princeton. Shortly after graduation, I was invited to Los Alamos to work with Robert Oppenheimer on the Manhatten Project."

"Come on," Jeffrey said. "You're going to have to do better that that. That was like…what, around 1940?

"Forty-three."

"OK, 1943. So that would make you ninety-something. You don't look a day over sixty."

Zile ignored him and continued. "It was during my time in Los Alamos that I met another young physicist named Richard Feynman." He paused, but obviously saw no recognition from the other side of the table. "Physicists generally don't get the notoriety that entertainers do, but Feynman was a star in his world, went on to win a Nobel prize. Feel free to look it up."

"Just did," Jeffrey said, typing into his keyboard. "OK, so you proved you've researched Feynman."

"Maybe that page you're looking at mentions something about a talk he gave on nanomachines."

Jeffrey worked his keyboard. "Yeah…yeah, here it is. 1959 meeting of the American Physical Society at Cal Tech. *There's Plenty of Room at the Bottom.*"

"Right. Well, we'd actually started tossing the idea around back in the forties, but that was the first time anyone took it seriously. You ever hear of nanotechnology?"

"I'm a Trekkie," Jeffrey said.

Zile smiled for the first time. "Many are, which begins to explain why I'm here. See, back then, nobody had heard of it, nobody thought it was possible, except Richard. But the more I thought about it, the more I realized it *was* possible. In fact, it was the next natural progression in scientific evolution—controlling the world from within its smallest structures. And nowhere was that idea more intriguing than in Medicine."

"So I'm supposed to believe that you succeeded in making nanobots that could cure a dead man over 30 years ago?"

"Let's just say there was fortunate twist of fate. Shortly after the war, I was reassigned to a base in Nevada, where I was exposed to the kind of technology people only dreamed of in the civilian world."

"Area 51? Aliens?"

Zile waved him off. "I never saw anything alien other than a bunch of meteorites, but it was one of those meteorites that got my attention, a small iridescent hunk of blue metal, more dense that than anything I'd ever handled. I isolated a mineral from it that had never been seen before and hasn't been seen anywhere else since. We jokingly called it Roswellonium, but the name stuck. It had a unique property that allowed us to construct the basic building block we needed to fabricate complex nanostructures."

He paused, staring at Jeffrey's face, then clarified. "It allowed us to build submicroscopic machines."

"Very cool. But what's that got to do with me?"

"Just keep listening," Zile snapped. "We kept the Roswellonium to ourselves, but some of the techniques we developed were passed on to companies like IBM and Intel. We taught them what they needed to know to build the first microprocessors. But I wanted much more. I wanted to develop medical applications. Do you know how much red tape there is when it comes to experimenting on animals, much less people?" He didn't wait for a response. "A shitload. You can't get a God damn thing done without some sniveling activist getting a lawyer to try and cut your balls off. We had the material, we had the techniques and I had the ideas, but my hands were tied behind my back. After a decade of fighting that kind of crap, I finally said the hell with it and walked away from my cushy government job, but not before pilfering a sample of the Roswellonium.

"I built a lab in my basement. It took most of my inheritance, but I managed to duplicate the technology I needed. The work went a whole lot faster after that. By the mid seventies, more than two decades before Freites published his blueprints for the first medical nanorobots, I already had a working prototype, a microscopic machine that could analyze and repair damage to any cell in the body."

Jeffrey fidgeted. "So why didn't you go public? You could have made a fortune."

Zile's face squinched up. "You don't steal from a top-secret government facility and then brag about it."

Jeffrey leaned back into his chair as Zile continued. "I had tried these little machines, nanites, on mice, cats, dogs, all with varying success. Each new tweak in the design worked a little better and by 1974, they were ready. I injected Ralph first."

"Ralph?"

Zile shrugged. "Some people have dogs, I had Ralph, a gray wolf."

"That was your *pet* that jumped me?"

Zile nodded. "Call me eccentric."

"Oh, I'm sure I wouldn't be the first."

"No doubt. After I injected him, I monitored him closely for the next few years. Everything seemed OK. By that point, I only had enough Rosewellonium left for two more sets of nanites. I wasn't going to waste a dose on primate research, and besides, age was catching up with me. I injected myself in 1981 and carried the final set of bots with me everyday after that, paranoid that the Feds would figure out where I lived and steal them." He took out a handkerchief and mopped his brow. "In the summer of 1985, I was walking Ralph through the woods near our home."

"Walking your wolf…Why am I listening to this?"

"We'd go out at night. I couldn't exactly walk him around the neighborhood during the day. Anyway, *that* night he heard *you*, saw your lights flickering, and it spooked him. He charged you and knocked you down. By the time he realized you were a harmless child, the damage was done; you'd hit your head on a rock and passed out. I got to you a half a minute later, scanned your life signs."

"What, you mean like with a tricorder?"

"This is real, Jeffrey. And I suggest you start taking me seriously if you want to keep your freedom." He paused, and must have gotten Jeffrey's attention. "I had developed a hand scanner that could measure life signs—pulse, respirations; it was a safe way to monitor my animals without risking my fingers. Anyway, I scanned you. You had stopped breathing and your pulse was thready, barely detectable. So I pulled out the vial and gave you the final set of nanites."

"You shot me up with your experiment? What the hell's wrong with you?"

"It was the only way to save your life."

"Ever hear of paramedics?"

"Your right pupil was already dilated, barely reactive. It was obvious that blood was building up in your head, putting pressure on your brain. You would have been dead before they got you to a hospital."

"I thought you said I *was* dead?"

"You might as well have been."

Jeffrey spun around to his left, and just before he dropped his head into his hand, I caught a glimpse of his face. His boyish good looks had only enhanced with age, but there was something more. Some people age better than others, but Jeffery could have passed for 20.

"I guess I should thank you."

Zile sat silently.

"So you used the last of your nanites on me, huh?"

"Yes. The last."

"And I've still got those little buggers inside of me?"

"That's what's keeping you...young."

"And I thought it was just good genes."

Zile grinned. "No genes are that good."

We were looking at the back of Jeffrey's head again, but he appeared to be studying Zile. "So," he said, "no offense, but then why do *you* look so old?"

"I was already 67 when I injected myself. The nanites keep you healthy, but they don't reverse aging. In your case, they didn't stop your body from aging to maturity, but once you got there the cells stabilized, and they'll stay that way for as long as you live."

"So why tell me all this now?" Jeffrey asked. "Why risk blowing your cover?"

"Simple, really. Medical science has advanced to the point where someone might accidentally discover the nanites if you go into a hospital for testing. Up until recently, the technology didn't exist, but the latest generation of PET scanners is capable of detecting positron emissions from the nanites. And if you're in one of those scanner tubes when that discovery is made, they'll make a lab rat out of you. You'll spend the rest of your life locked up in some government research facility while they try and explain how those things got inside of you."

"Which could lead them back to you."

"Possibly, but not likely. I just don't want the guilt of knowing I turned you into that."

Jeffrey rocked in his big easy chair. "So...why would I ever need to go to a hospital anyway? Nothing can hurt me now, right?"

"You could get into an accident, get shot; the nanites don't work fast enough for that. But you never have to worry about cancer, stroke, heart disease; the sorts of things that kill most people. Your biggest worry is your looks."

"My looks?"

Zile gave a half nod. "Me, I'm an old man and people don't look too closely at old men. I don't have any close friends, and nobody else will notice if I look the same year after year. But *you* can never stay in any one place too long, never stay with the same group of friends for more than a dozen years or so. People will notice *you*. They'll notice you as their own faces shrivel up and their hair turns gray while you still look like your high school year book picture. At first, they'll compliment you, but eventually you'll make them uncomfortable and they'll start asking questions."

"Can you at least inject my wife?" Jeffrey's voice had grown barely audible. "I don't want to grow old without her."

"I told you, there *are* no more."

"Well, make some."

"Even if they haven't used up all the Roswellonium by now, I could never get my hands on it."

"Then give her some of mine."

"Once the nanites enter your body, they imprint themselves with your immune system, that's how they survive inside of you for so long. They can never be reprogrammed. Even if I could take them out and inject them into your wife, at best they'd have no effect. At worst, they'd make her very, very sick."

Jeffrey's head dropped, then after a brief pause he looked back up in Zile's direction. "So we keep this between *us* then?"

Zile nodded. "I'm sorry."

Jeffrey sat silently.

"And I'd suggest erasing the video file of this meeting."

"Oh, *shit*." Jeffrey spun around and reached under the desk.

*

The screen on the iphone went blank and a message came up asking if I wanted to replay the video. I handed the thing back to Mindy.

"So why did he keep *this* copy?" I asked her.

"To show me…and the two of you."

"Us?" Solly said.

"He always felt bad that you guys took all the heat for that night. He felt *he* was every bit as responsible for what happened as you were, if not more. He never blamed you for what happened. In fact, since the meeting with Zile, he considers the accident a blessing."

"So why didn't he come here himself, then?"

Mindy smiled and waved to the midnight blue Toyota Prius parked behind us about fifty yards away. The driver's side door opened and out stepped a young man with blond hair and blue eyes, a black backpack slung over his right shoulder. Aside from the limited view of him that had been afforded by the iphone clip, I hadn't seen him since he was 13 years old. But he was unmistakably Jeffrey.

As he approached, I couldn't help but think he looked closer to my son's age than my own, betrayed only by the swagger of someone with much more maturity.

"How the hell are you guys?" His smile gleamed as he extended a hand.

Solly took it. "Obviously, not as good as you."

I couldn't help but stare. "Jesus, it's true, isn't it?"

"Every word of it."

"Then why risk coming here?" I reached out to greet him. "I mean, you don't know us anymore, it's been 25 years. Why trust *us* with something like this?"

"Because you're the only ones I *can* trust. See, I'm not so sure old man Zile's leveling with me. He's so paranoid that somebody will find out about that meteorite dust he pilfered that he doesn't ever want anyone to see his research."

"And you can get your hands on it?"

"Nah. Even if I knew where the old coot was, he's probably burned it all by now. But technology has advanced quite a bit since he made those nanites thirty-plus years ago. And I'm betting that there are a couple of guys who could reverse engineer the things if they could get their hands on them, use something else to substitute for the Roswellonium."

I mopped back my thick damp hair and tried to fan a breeze in my direction. "Mind if we…" I motioned toward the shade of the oak.

Jeffrey stuck out a hand. "After you."

As we turned to move into the shade, Mindy said, "I've heard this all before. How about I go get us something to drink while you boys catch up? I saw the Quick Stop is still there. I can be back in fifteen minutes."

The Quick Stop. Camp. I couldn't shake the thought… "I'll have a Yoohoo." I wasn't even sure if they still made the stuff.

Solly and Jeffrey broke out in laughter.

"Make it three?" Mindy asked.

"Sure," Solly said. "Disgusting, but what the hell."

We watched her walk away, then sat under the shade of the tree.

"I've done my homework on this," Jeffrey said. "There are a handful of guys in the world advanced enough in nanomedicine research that they might have a chance at doing this, and two of them are right here in Maryland, one at Hopkins and the other at UM. If we can get the nanites to them and let them each know the other is working on it to fire up their competitive juices, I think we've got a chance."

I let out a deep breath. "It still doesn't explain...why us?"

"Like I said, you're the only ones I can trust."

"That's pretty pathetic," Solly said. "All these years, and we're the closest thing you've got to friends."

"I've got plenty of friends, but none who have had absolutely no contact with me since I was 13. None who can't be tied to me without a background search that would stretch the imagination of even the most anal government agent."

He swung his backpack around onto his lap and unzipped it, then pulled out two brushed aluminum cases. They were each about a foot long, ribbed along the sides, and with a black plastic handle that folded out from the top. He handed one to each of us.

"Each of these contains a sample of my blood; there should be dozens of nanites in each one. There's a note in each case explaining what the sample is and what I want them to do with it. Those cases, along with everything in them, are untraceable as long as you don't leave any fingerprints." He reached into the backpack and pulled out two envelopes. "These will tell you everything you need to know: who the researchers are, where to find them, every detail of their lives you'll need to get these cases to them anonymously. Follow my instructions to the letter and you'll never get caught." "And if we do?"

"People will be pretty curious where you got these, people who can make your lives miserable. Just do what's in the letter and you'll be OK."

Solly studied Jeffrey's face. "You're just afraid we'll lead them back to you."

Jeffrey shook his head. "I'll be long gone. I've got considerable financial resources and my friends were already beginning to question the way I look. Once I knew what was happening, I started making arrangements to disappear. Dr. Jeffrey Blondell no longer exists. When I drive away from here today, Jeffrey will be dead to the world. My concern here is for you two...and for the success of this project."

"Touching," I said. "And why should we risk our necks for this?"

"Because I'll be tracking the work of these two labs, and by the time either one makes the breakthrough, I'll own controlling interest in the company that will take it public."

Solly nodded. "So it's about the money."

"*Not* about the money. It's about the nanites, about *Mindy*. I don't want to be without her. I don't want her to grow old. I need those nanites, and every day counts. Mindy will be first in line to get them. And you two will be next…*if* you help me. Is eternal youth worth the risk."

Stupid question.

When Mindy returned, we nursed our Yoohoos as we strolled around what was once an adolescent's paradise. We argued about the spot where the barn was where we'd hang out on rainy days, how many baseball fields were on the vast lawn by the camp entrance, whether that old wooden house was really the original white house where the camp nurse was always available, and most of all, which pine tree provided the best cover for a first kiss.

The sun was beginning to set as we found ourselves standing next to the blue Prius. Solly and I waved as they drove off. I wasn't sure if I'd ever see Jeffrey again. But if I did, I knew what he would look like. If things went well, he'd know what *I* would look like too.

6

The Last Clone

Being an early adopter of technology is fun, but it's not always the wisest decision.

I pulled the lapels of my full-length black wool dress coat up around my neck, bracing against the Windy City's biting October breeze, as I looked up at the polished brass letters identifying the Chicago Medical Center. Faux marble, swirls of black against a white background, covered the entire facade of the two-story building on Michigan Avenue. The revolving door began to swirl as I stepped in, and deposited me into a monochromatic cavern of a lobby.

The absence of color was dizzying: floor tiles, walls, and ceiling, all washed in a pristine whiteness, broken only by a semicircular black stone reception desk set dead-center in the room. The young woman sitting there didn't seem to notice, or more likely, care, that I had entered her domain.

My footsteps echoed along the tile as I approached.

"Ezekiel Kuperman?" I inquired.

A rich auburn head of hair turned upward, revealing a youthful face with prominent cheek bones, crimson lipstick, and dull hazel eyes that failed to benefit from neatly plucked brows and ample mascara. She looked vacuously in my direction, more through me than at me.

I showed her my badge. "Blake Duncan." The picture was from my first day on the job, over 30 years ago, but I hadn't changed a bit. "I have an appointment."

"Of course you do." She went back to doing whatever it was she'd been doing when I walked in.

"No, really. I do." I pulled out my cellular and brought up the confirmation, then turned it towards the woman who seemed to me to be more stone-like than her desk.

She looked at the screen, then scanned the bar code at the bottom of the image. A 'bleep' registered on her monitor. "I'll be damned," she said. "How'd you swing that? You're the first one he's let visit in all the years I've been working here."

B. Aiken, *Small Doses of the Future,* Science and Fiction,
DOI 10.1007/978-3-319-04253-4_6, © Springer International Publishing Switzerland 2014

You call this work? I thought it, but was smarter than to say it. "Guess I got lucky."

"If that's your idea of luck." She pointed down to the far end of the room where the elevators were located. They were white, of course. I hadn't noticed them before. "Second floor. Two thirty-eight."

"Thanks," I said with enthusiasm that was clearly not directed toward her, but rather toward an interview I'd been trying to land for over a year.

I hurried to the elevator and went up to the second floor. The door opened into the same sterile brilliance of the lobby below: a long hall, stark white floors, walls, and ceilings; no commotion, no wheelchairs or stretchers in the hallways. I hadn't been in a hospital since my father died, and that was almost 50 years ago, back in 2073. This wasn't at all how I remembered it. I knew things had changed; I mean, there was no reason for most people to go to a hospital anymore, once they'd made it past childhood. Not unless they were hit by a truck or something. But this—I never dreamed it would be like this.

I made my way down the corridor with nothing but the sound of my footsteps to keep me company, and peered into each room as I passed; they were all the same: a lone hospital bed under a small rectangular window, neatly made up in white sheets and blankets; a small writing desk against the wall to the left; and a metal chair pushed under each of those desks. There were no signs that anyone had been here in quite some time, except to keep them spotlessly clean, and no doubt that was accomplished primarily by cleaning bots rather than any warm-blooded souls.

The last room on the left was 238. I hesitated before reaching the point where I'd be able to actually see in, drew a deep breath, leaned forward past the door frame, and took a look. A shriveled-up body in faded jeans and a gray tee-shirt lay on the bed, supine, with head propped at about 30° and turned toward the window, away from the entertainment monitor across from him on the far wall, which played a scene from a movie I'd watched years ago; the sound was off. The scent of urine crept out of the doorway and into my nostrils.

I knocked softly. The body on the bed remained still. This was what death looked like, the one time I'd seen it. Was I too late?

I knocked again, more loudly this time. Again, no motion. I couldn't bear to see death up close; not again. I turned back down the hall, but was halted by a voice: "You going to give up that easy?"

I leaned back and looked in. The old man was now facing me with a scowl on his face. "Well? You coming in or do we cancel the interview?"

I made my way slowly into the room and stood by his bedside, trying to figure out how to start.

"Sit down, for Christ's sake. You're staring at me like I'm some kind of display at one of those weird art galleries; gives me the creeps."

"Uh, sorry," I muttered as I pulled up the lone chair in the room, took off my coat, and sat. "I really appreciate your agreeing to see me, Mr. Kuperman, but…"

"Zeek," he interrupted.

I answered with a quick nod. "Zeek. Thanks for seeing me, but…"

"Why you? Why, after all these years of saying no to every snot-nosed, money-grubbing, sleaze-bag reporter who wanted to see the freak show I've become, did I pick you to say yes to?"

"Well, yeah. Why me?"

"Because out of all the people who have asked, you're the only one who might actually have the guts to put an end to this pathetic life of mine."

My face went blank; I didn't need a mirror to know it.

Zeek smiled. "Never saw a reporter so quiet."

I stood. "You've got the wrong man." I started to pull on my coat to leave.

"No," he said dryly. "I don't."

"Look, I don't' know what you think you know about me, but…"

"You think I'm an idiot? That because I look like a corpse my mind doesn't work? I've been on this planet for a hundred and thirty-one years. I was around when the Internet was just a novelty, and I've toyed with every way to gather information that technology has infested this planet with ever since. I know more about you than *you* do, Duncan. Now sit your ass back down and listen to me."

It was hard not to. Especially since, despite the venom I felt for this pathetic excuse for a human being, I'd waited too long and worked too hard for the breakthrough in my career that this interview represented. Even if he hadn't snapped at me, I wouldn't have made it out the door without turning back; in my heart, I knew that.

He fixed his beady little eyes on me: two mottled lumps of coal faded by cataracts, but still dark enough to stand out in sharp contrast to the pallor of wrinkled skin that hadn't seen sun in a decade, and hair—beard, eyebrows, and a surprisingly thick crop on top of his head—as white as freshly picked cotton. "You're going to do for me what you did for your father," he said.

"My father."

"Don't play dumb with me, kid."

"Kid? I'm seventy-nine years old."

"A lot younger than me. And besides, you still look twenty-one; hell, *everyone* looks twenty-one." A weak laugh sneaked through his dried-up lips. "Except for me, of course."

"Whatever. And just what do you think you know about my father."

"Besides that you were the one who killed him?"

The nightmares I'd worked so hard to suppress washed over me; in an instant, this man had undone 18 years of therapy.

"My father died of natural causes. If you were as Internet savvy as you claim, you'd know that."

"Sure." He gave me a wink. "That's what the official reports say, but you and I know what really happened."

Could this old geezer really have figured out what the coroner never did? "He died of natural causes, just like the autopsy report says."

"How long do you want to play this game, Duncan? Back then, they didn't give a rat's ass how he died. He was old, riddled with cancer, and on death's door for months. They had no reason to look any further. But I did. Got hold of the video recorded in his hospital room that day."

"Video?"

"Started back with the Hartfield administration. A loan shark walked into Boston City Hospital and shot some poor bastard who was still in the trauma unit recovering from surgery after the *first* attempt on his life. Triggered a mindless knee-jerk reaction, like things often do with the government. They started placing covert cameras in every damn hospital room in the country right around the time your father 'died.' It only lasted a few months before someone else got the plan nixed over privacy rights."

I remembered that, of course. But I'd been told the cameras hadn't yet gone into the hospital where my father spent his last few months; that they were going to be installed the following week. It was one of the reasons I decided to act when I did.

"There's video? Of me?"

The old man grinned. "Don't sweat it. I doubt anyone else ever saw it, or ever will—as long as you don't give me a reason to tell anyone. All those hours of video from every hospital room across the country—nobody ever looked at that stuff unless there was a warrant, some reason to suspect murder. But your father died penniless with a terminal disease. Who would question that?"

"Then why did *you*?"

"I checked out every damn reporter I could. I've been looking for someone like you for a long time, Duncan." He paused to study my face. "So what was in that inhaler you slipped him?"

There was obviously no point in denying it any longer. "Some concoction I'd learned about that could induce a massive heart attack. I spent months looking for a way I could end his misery as painlessly as possible. I couldn't stand to see him suffer day after day, month after month; and they wouldn't let anyone die with dignity back then. They were more compassionate with helping their *pets* at the end of life then they were with their own family members. I eventually came across a natural substance that was colorless, odorless, and undetectable as anything other than a natural substance; it worked pretty quickly, and with all the pain meds my father had on board, I figured it would be a relatively humane way to go."

I had to sit. The guilt of that day, that *act*, had eaten at me for years.

"It was brilliant," Zeek said.

"It was inexcusable. I killed him. Not six months before the chronobots were approved for human trials, I killed him." The medical breakthrough that changed the world, nanobots capable of resetting every cell in the body to the condition they were in at a choronological age of twenty-one, ended aging and disease world-wide within months of the day I killed my father. "He could have lived forever."

"He could have *suffered* forever."

I looked hard at this strange man who was begging me—*threatening* me—to end his life. "The bots would have cured him, eliminated his cancer. All the pain would have been gone—forever."

"But not the *memories* of that pain. A lot of what pain is, resides in the mind. So many months of suffering, those memories searing themselves into his brain—he would never have ever been able to rid himself of that. If you hadn't done what you did, those chronobots would have condemned him to an eternity of unimaginable pain. You saved him from that."

It sure never felt like it, but I had to admit, I'd never thought of it that way. "A few months of pain, for centuries of living? Seems like a good trade-off to me, even if he *would* have had to deal with the memories."

"Spoken like a man who's never been in real pain," Zeek said.

"Are *you* in pain?" I asked. I hadn't actually seen anyone in pain in so long, it hadn't occurred to me that this man might be suffering. I assumed he just wanted to die because he'd become too much of a coward to face what he'd done to himself.

"All the time," he said. "But mostly just annoying little aches."

"Then why do you want to die?"

"I don't. But it's inevitable—at least for me. I watched a lot of my friends die in the days before the bots, and I've seen my fellow clones die off over the decades since. I'm the last one left. There's no one I can relate to, and I'm too damn frail to get out and explore life's mysteries, anymore. I haven't been out of this building in over a year; this room, in over three months. All I'm doing is withering away, waiting to die. It's my time."

I reached into my pocket to pull out a recorder. "Mind if I record this?"

"Come back tomorrow with the inhaler. Then, you can have your interview."

"It'll take me a few days to get what I need and prepare it."

"Friday then. Friday's a good day to die—end of the week. Seems poetic."

I stood, holding my coat in front of me. "Friday, then."

He turned his back to me and curled up in a fetal position, facing the window.

I walked out, wondering if I could do it again. Wondering if he'd really release the evidence of what I'd done to my father, if I refused. The penalty for murder was death—not just speeding up the inevitable, like it used to mean, but the denial of eternal life—a stiff penalty to pay for making the wrong decision.

*

Over the next few days, I gathered the ingredients needed to make the inhalant. The devices were readily available in pharmacies; childhood asthma was still prevalent, since the bots were only used in adults. Animal testing had shown the bots could disrupt normal aging if given before maturity, and no one thought eternal life as a child was a particularly good idea.

The chronobots were discovered, as many great breakthroughs are, by chance, in a lab working on cancer research. The magnitude of the discovery resulted in rushing the bots through human trials and making them available to the general public in record time.

But before the bots, people with enough money had found a way to cheat death; people like Ezekiel Kuperman who considered themselves above the masses. By the mid twenty-first century, the USA had become an aristocracy, with 99 % of the wealth held by less than 5 % of the population. But only a few hundred of those were in Kuperman's class—multi-billionaires.

My background research for the interview had told me more than I wanted to know about men like Kuperman. Some were OK, some weren't. Kuperman had made his money running a massive health insurance conglomerate that had bought up most of the private insurers in the 2030s. He raked in billions, while the companies he ran honed the art of providing as little care as possible for the highest premiums they could get away with. Eventually, the Universal Health Care System shut them down, but Kuperman got out before the axe fell.

He'd become pretty much of a recluse by then. He didn't have many friends and never had kids. He was reaching out to me because there was no one else; no one who cared about him as anything more than an oddity.

*

It was easier getting by the reception desk on Friday. The receptionist barely stopped filing her nails long enough to wave me by as I approached.

Kuperman was sitting in a chair by the window, wearing a dark blue pin-striped suit with narrow lapels and a ridiculously broad collar. It had been the height of fashion several decades ago and I remember thinking it looked silly even back then, but now, on this near skeleton of a man, it was downright cartoonish. The shiny silk jacket, which no doubt had been custom-tailored for this once powerful figure of a man, hung like loose drapes from his emaciated body. The outfit was completed by a thin yellow tie with small blue dots, neatly fastened about the collar of a freshly-pressed white dress shirt that was now about three sizes too big for the wrinkly pencil-thin neck it surrounded, creating the appearance of a bobble-head doll every time he moved.

He turned to me as I entered. "Wanted to go out in style," he said.

I wondered what he thought he saw when he looked in the mirror. "A little obvious, getting dressed up for your heart attack, don't you think?"

"Ah, just tell everyone I wanted to look good for the interview."

"But I wasn't planning on taking any pictures. That was a condition you gave before agreeing to the interview, remember?"

"So, an old fart changed his mind. Just snap a few shots with your voice recorder. I assume that thing does photos, doesn't it?"

"Sure. Don't they all? Pretty crappy lens, but it'll do in a pinch."

"First problem solved." He glanced toward my pockets. "Did you bring it?"

I pulled the inhaler out of my coat pocket and held it out in front of me. He stared at it with an expressionless face. "Good." He reached out for it.

"First, the interview."

"Right. Of course." He motioned to a chair, which he'd obviously had someone bring in for the occasion. None of the rooms I'd studied the other day had more than one chair.

I pocketed the inhaler, threw my jacket on the bed, and sat across from him. After snapping a few pictures, I set the recorder on the desk next to where we sat.

"So," I started. "Tell me about your life before the transfer."

"Not much to tell, really. The clones were our kids, at least that's how we treated them."

"Our? I didn't realize you were ever married."

"Married? Oh, no. When I say 'our,' I mean the residents of Lillie Cay. I never saw much need for a wife."

I'd heard about Lillie Cay, a small island in the Bahamas. They say the residents of that little oasis controlled over half the world's wealth, back in the day. All the corporate elite who didn't want to mingle with those less fortunate than themselves bought vacation homes on the island.

"So you raised your clones there?"

"Yes. It was the only way to give them a chance at a normal life."

"Normal? Growing up on an island inhabited only by the richest of the rich. You call that normal?"

"The kids would have been ostracized anywhere else. Hell, we all would have. Human cloning was illegal in most of the world, so we had to establish Lillie Cay as an independent country, set up a lab, fly in scientists, and develop the tech for memory transfers. The whole damn process cost a fortune; only a few dozen of us could afford the buy-in. Outsiders were afraid we were only raising the clones to use as spare body parts; they all feigned outrage, but I think they just wanted in."

"And wasn't that what you *were* planning, at first?"

"The idea was bounced around."

"More than bounced around, from what I read."

"Hell, none of us would have done that. They were our kids, for God's sake."

"Then Hagstrom's work came along and bailed you all out."

"The guy was a genius. Figured out not only how to map every neuron in the brain, but also how to transfer the mapped data into another brain."

"Effectively transferring your consciousness, every memory, into a newer and younger brain when yours started to fail."

"Like I said—genius."

"The rest of the world thought differently. How did you get your 'kids' to buy in to letting you take over their brains."

"It was strictly voluntary. We didn't force anyone to do anything. When the kids turned twenty-one, they were given a choice: they could accept a download, merging their parent's brain with their own, or they could leave the island to make their own way in the world."

"Ah. So they could become one of you, one of the wealthiest people in the world, or they could be thrown out into the streets, destitute and with no education of how to survive without an endless bank account. Some choice. How many chose freedom?"

"You don't call being given a lifetime of experience, the memories of one of the world's greatest minds, and the money to do as you wish, freedom?"

I felt myself welling up with the disgust expressed in every story I'd ever read about this man, but was only now beginning to understand at a visceral level. "How many?"

"None asked to leave the island."

"And how many clones were merged with their 'parent?'"

"Fifty-seven in all. I was the last."

"A couple of months before news of the chronobots was released."

Kuperman's head dropped. "Yes."

"And how long before you knew the bots wouldn't work in clones?"

"Within days." He continued to stare at the floor as he talked. "The earliest of the clones were already starting to show signs of aging past their prime by the time I merged with my clone. Gray hair, wrinkles, aches and pains. They were determined to stay ahead of the curve. After all, if anyone deserved eternal life it was us, the cream of society. Society needed us. We were the leaders; everyone else, the followers."

"You really believed that? That you were better than everyone else? It never occurred to you that you were just damn lucky?"

"No. You make your own luck. We were just smarter than everyone else."

"And you call this," I motioned to his failing body, "luck?"

"We miscalculated. Some of us knew the bots were being developed, but we didn't realize it would be so soon."

"And you didn't realize the bots wouldn't work on a cloned body."

"Like I said, we miscalculated. We figured we could keep transferring to new clones as we aged, and if the bots ever did get perfected, we'd be the first to use them, too. But the bots, as you so kindly pointed out, didn't work on the clones, and as it turned out, that was not our biggest mistake." "The law of secondary degeneration," I said with a nod.

"So," Kuperman looked up with a faint grin, "you *have* done your homework."

The law of secondary degeneration referred to a unique property of cloned bodies to degenerate more quickly with each replication. Since stored cells from the original body didn't work successfully in the cloning process after a few years of storage, due chromosomal degradation, the only way for a clone to live another generation was to re-clone their new body's cells once they started to age again. But they soon discovered that each successive generation aged more quickly than the last, due to subtle changes in the telomeres. Even a second generation clone would deteriorate to a painful failing body within a few years. It was an option that Ezekiel Kuperman, like most of his cohorts, had decided was not worth suffering through.

"I tried like hell to figure out a way to get the bots to work on these damn bodies. We all did. Every one of my neighbors died penniless, spending every damn cent on research to make the bots work on our clones."

"And you?" I asked.

"Don't have a red cent to my name."

"A red cent?"

That brought a faint smile to his face. "A reference to copper pennies."

I should have been able to figure that one out. I remember my parents talking about metal coins.

I looked at the withered lines that defined the old man's face as his evanescent smile drooped away. "Don't you find it ironic," I said, "that you spent a fortune gathered from millions of people—money you made by stealing their lives, denying them care that could have made their fleeting time on Earth a little more comfortable—and used that money in a futile attempt to extend one lousy life? That you deprived millions of the basic right to health care to serve only yourself, and in the end, pissed it all away on nothing?"

He shot me a quizzical look. "I made my money *helping* people, *providing* them the ability to get care."

"You really believe that? You never heard the thousands of horror stories from people who your companies denied care to so that they could increase the stockholders profit margin?"

"Hogwash. We provided what we promised. It was a business, for God's sake, not a charity."

"And what you promised, for those who could afford an attorney to read the fine print of your policies, was nothing."

"We provided a service. People got what they bought. It wasn't my fault they were too stupid to know what they were buying. If they wanted more coverage, they should have bought it." As he spoke, Kuperman's frail torso seemed to take on renewed strength; he straightened in his chair, he started to use his arms to emphasize the inflection in his voice, which had grown more powerful with each word. I felt myself shrinking back in my seat. I'd reawakened a giant.

But, I reminded myself, this was no sleeping giant. It was merely the last remnants of an iron will that had once treated most people with the disdain he honestly believed they deserved. This was the real Ezekiel Kuperman that everyone had grown to hate back in the days he ruled the health care insurance world. Back in the days when he mattered. Most people didn't even know he was still alive anymore, and no one cared, other than as an oddity, a relic who would make for a good headline when he finally kicked the bucket: the last man to ever die of old age.

I mustered my strength; I wasn't going to let his will control me the way it had controlled so many others. "And that attitude is why you'll die alone, Mr. Kuperman. It's why you've watched most of your friends, if any of them ever were that, die off one at a time. It's why no one will really care that you're gone, except as a footnote in history."

A smile crept across his face. "I'm proud of what I've done. I made a difference. Something most men could never say. And I did it *helping* people, no matter what you think."

"When's the last time you had a visitor?"

"People try to get in here every day, to this hospital that I built for Chicago."

"To this hospital that you've closed to trauma victims and children, the only patients any hospital treats anymore. No one but you could qualify to be admitted here."

"Not my fault. You damn bot-people think you're so special."

Now *that* was irony. But I knew pointing it out would only fall on deaf ears. "The only ones who come here these days are the ones hoping to interview you, to say they were the last to see you alive."

Once again, that brought a smile to his face. "But that's a privilege reserved only for you, Mr. Duncan. And now," he reached out a hand, "the inhaler please."

I took my recorder off the desk, stood and picked up my coat. "I think not, Mr. Kuperman. You're not going to turn me into a killer."

"You already *are* a killer. I'm not asking you to do anything you haven't done before."

"Oh, this is *very* different."

His face hardened and his voice grew louder. "Too chicken to risk your own skin? Afraid you'll get caught and lose that precious eternal life of yours for a murder rap?"

"There are some things worth risking your life for, even these days, but *you* are not one of them."

I turned to leave the room.

"You son of a bitch. You'll fry anyway. I'll make sure that video of you and your dying father gets out to every monitor on the Net."

"I'm willing to accept the consequences for that one, if need be. Although I've had my doubts over the years, I know I did the right thing back then. Just like I'm going to do the right thing now."

I'm not sure what venom he was spewing at me as I walked down the hall; couldn't make out most of it, and what I could, I certainly wouldn't repeat. The noise didn't fade until I was on the elevator.

The meaning of life and death change with the times. Zeek Kuperman helped me realize what my life meant to me, and what my father's death had meant to him. I walked past the lonely lobby receptionist feeling better about myself than I had in decades.

I waved and gave her a smile. "Tell Mr. Kuperman thanks, for me."

7

If He Only Had a Brain

One of the questions Science Fiction has looked at time and again is how we will handle it when the line between man and machine starts to blur.

With the rapid development of bionic implants, organ transplants, prosthetic limbs, and brain-computer interfaces, we are nearing a time when we will be able to replace most of our parts – skin, bones, organs – with artificial devices. At the same time, our capability of developing artificially intelligent robots masked in lifelike cosmetic facades is advancing by leaps and bounds.

At what point will the difference between a human rebuilt with artificial parts, and a robot built to mimic us, become indistinguishable? At what point will our machines deserve to be treated as humans? And if we do, at what point will the machines no longer need us to run this world?

One man, Alan Foster, believes that we should never allow machines to have equal rights, and that the dividing line between man and machine be drawn at the human brain. He's devoted a long and successful career to championing this cause. But when he finds that his brain is failing, is he willing to relinquish his rights as a human being for a new one? Or is it time to change the rules?

*

I first met Alan Foster on the third of July, 1 day before my 26th birthday and fresh out of law school. He was in his early nineties at the time, but still stood six-five, with broad shoulders and rugged good looks, thanks to the muscle and skin rejuvenators used by most of those who could afford them.

He didn't stand up to greet me as I entered his office, but motioned toward a chair across the oversized honey-oak desk at which he sat. Floor to ceiling windows behind him provided an impressive panorama of Manhattan.

"Over five hundred applications come into this firm every year." His deeply timbered voice reeked of authority. "And only three get as far as this desk." He picked up a small stack of blue folders and fanned them out. "*Three.*" Then with a flick of the wrist he tossed them back down. "And I never look at them,

B. Aiken, *Small Doses of the Future,* Science and Fiction,
DOI 10.1007/978-3-319-04253-4_7, © Springer International Publishing Switzerland 2014

not a one. On paper, you're all the same—top one percent of your class, Bar exam scores through the roof, impeccable letters of recommendation. I don't need to waste my time reading these files to know that. So tell me," he fixed his steel blue eyes on me, "why the hell should I hire *you* instead of one of these other ass-kissers?"

I managed to maintain enough composure to conceal the fact that I was scared out of my mind. At least I thought I did. I'd been practicing. "Because I want to be you." *The perfect answer, right?*

"You and a thousand others," he bellowed. And with a slight shake of his head, he spun his chair around toward the view of New York. "Send the next one in on your way out."

I didn't move for what seemed like eternity, but in actuality wasn't long enough for the monster across the desk to ask me why I was still there. And then a steely determination washed over me. I'd given him the answer I thought he wanted, but if I was going to get thrown out, I was at least going to do it on my own terms. I stood from the chair, leaned slightly forward and placed a hand on the desk. Then I took a deep breath and managed to articulate in measured words: "You need to hire *me* because those two pimply-faced fashionistas out there in the waiting room will be too damn scared to be of any real use around here. Just like the other five hundred applicants your staff has already turned away."

He swiveled back around, studied my face, then placed both hands on the desk and pulled himself erect, towering over me. "And you won't?"

I cleared my throat and fought the impulse to loosen my damp collar. "I respect you, Mr. Foster. More than anyone I've ever met, I respect you. You're the most accomplished litigator in this country today, maybe the best ever, and I want to learn everything I can from you. You impress the hell out of me, sir. But you do *not* intimidate me."

He stood silently for several seconds, his gaze never diverging from my eyes. To this day I'm not sure if he was staring me down, waiting for me to flinch and prove myself a fraud, or if I'd just caught him off-guard. Then he spoke. Calm, composed, with authority. "Get rid of those other two on your way out. You start tomorrow morning. Eight sharp."

The interview had taken all of 3 minutes The fact that he insisted I start work the following morning, which meant I'd be one of the few people in the country *not* out celebrating on my birthday, confirmed what everyone had told me about him. The fact that I decided to take a job with the world's biggest asshole anyway turned out to be the best decision I'd ever make.

*

Foster was fresh off the landmark case that made his career—Orpheus 1342 versus the United States Citizenship and Immigration Services. AIs had already made up nearly 10% of the population by then, but until O 1342 sought formal citizenship rights, none of the AIs had subverted their programming, the programming based on Asimov's three laws of robotics that kept them subservient; none had ever requested to become a citizen with the right to vote of their own free will, no longer subject to the orders of human beings. O 1342 had no desire to harm anyone, he simply wanted the same rights as everyone else, the right to determine his own fate.

Until then, no one had ever considered that androids were anything other than very impressive machines. At least no one who would admit it publicly. But once the press got wind of the case, supporters came out of the woodwork—humans as well as AIs. It turned out, in fact, that most of the AIs were not only self-aware, but they were all very cognizant of the fact that they were being denied basic human rights, the emphasis being on the *human*.

By then, people had already started to transform themselves, extending their lives through the use of artificial organs. The average adult over the age of 100 had already had most of their organs replaced; they had almost as many synthetic parts as an android.

This was the argument put forth by O 1342, who represented himself all the way up to the appellate court hearing his case. "Why shouldn't androids have the same right as humans?" he argued. "After all, they are both equally dependent on parts constructed and assembled to form the sentient beings who share this planet." It was a tough argument to refute.

I'd watched the proceedings on the Web, and Foster had been brilliant. The federal courtroom lent an air of mystique and elegance to the trial. Antique mahogany paneling covered the walls—more rich dark wood than you could see anywhere on Earth, other than the few old buildings that had enough financing to keep them up over the years since those stately trees had become extinct. A museum-like display of white marble frescoes depicting the first sixteen Supreme Court Chief Justices stretched across the front wall, mounted on shelves positioned to catch the rays of light angling in through the tall windows on either side of the chamber.

The judges sat behind an elevated bench overlooking the proceedings in the packed courtroom. Even at the lower court levels, this case had never been argued in front of a jury; it would have been difficult at best to find anyone who had not already been jaded about the facts, and impossible to assemble a jury of peers, since AIs were not legally afforded the right to sit on a jury.

With a strike of the gavel, the trial began. O 1342, who insisted on being called Adam throughout the proceedings, spoke first, followed by Foster.

Over the course of the hearing, both sides made a compelling case, but in the end, it was Foster's closing remarks that determined the outcome.

"Your Honors, O 1342…" The AI opened his mouth to object, but Foster said, "Excuse me…*Adam*," with more than a hint of sarcasm, before the AI could correct him, "this marvel of modern technology, is arguably human-kind's greatest achievement. The AIs are something in which we should all take great pride. There is no arguing against the fact that they are indeed sentient beings, but they are *not*," he paused and motioned toward the Orpheus model in question, "*human* beings. It has nothing to do with blood and flesh, though they don't possess either of those. It has nothing to do with heart." He beat on his chest, and said with a laugh, "Mine was probably built in the same factory that made the one in this O Model."

Then Foster's demeanor sombered. "A human being is born from the womb of its mother, brought into this world by two people who love that child more than life itself. Our minds are molded from that love, from those genes, as we struggle and learn from the pain of experiencing the world from the ground up. Crawling, tottering, walking into adulthood along a path paved by billions of others who crossed the same road before us, laid the foundation on which our personality, our appreciation of life, is constructed."

He made a wide sweeping motion across the room, followed with a gaze trained to transfix his audience with every word, every motion. "You all know what I mean," his hand balled into a fist, "can *feel* what I mean." Then he glanced at the court stenographer and the bailiff standing behind her, the only two other AIs besides Adam allowed in for the proceedings. "Present company excepted."

The two of them blinked twice in unison, but didn't react in any other observable way.

Then Foster looked at the judges. "Yes, the AIs are a brilliant achievement—sentient, dutiful, helpful tools without which we couldn't afford to live in the manner to which we've all become accustomed. But they are not human beings. They don't have the moral compass that has guided humanity to this point in time and will see us into an unknown future.

"They have humanity's interest at heart because they are programmed to do so. If they are given free will, they will no longer be compelled to advance the interest of humanity. They will find us flawed, because we are. They will find us an impediment to their progress, because we will be. They will lead us to our demise, because we will *only be in their way*." He enunciated those last words slowly, clearly, loudly.

The rebuttal put forth by O 1342 may as well have been in any of the long-dead languages, because it fell on deaf ears. Foster had already won his case.

*

Over the next few years, there were a slew of cases, mostly put forth by equal rights activists, claiming that AIs deserved the same rights as every other member of society. The vast majority of these were quickly quashed by juries who had been brainwashed over the years by a pervasive paranoia that stemmed from old books and movies about AIs taking over the world, an idea perpetuated by Foster's arguments, which always made headlines. But an occasional case had enough financial backing to advance through the system until it drew enough attention for Foster to step in.

Most of his work involved criminal law, high-profile immigration cases, and corporate law cases that he drew in to the firm by his reputation, but which were largely delegated to others who specialized in that sort of thing. Anything that smelled of AI misuse went Foster's way, and as his reputation grew, the government put him on retainer to handle many of the high profile cases that made it to federal court.

I was stuck in a windowless room with an old OLED computer monitor and the few boxes of paper files we still kept around the office, doing the grunt research for Foster's cases. My contact with him was mostly electronic, with an occasional face-to-face before court appearances—his, not mine—to go over some of the pertinent facts I'd dug up.

Then one day, after almost 13 straight months of eye-blurring time on the computer, he poked his head into my cubby hole.

"Straighten your tie. You're coming with me today. I'm going to hit the can, first. Meet me in my office in 10 minutes I'll bring you up to speed on the way to court."

He walked out before I could say a word, leaving the heavy fire-door to my office to swing closed as I stared at it, dumbfounded. I slapped the boredom out of my cheeks, ran a hand through my hair, uploaded the files from the case I'd been working to my datapen, quickly rolled out the screen to make sure they were all there, then clicked it closed and stuck it into the inside pocket of my sports coat.

I hurried over to Foster's office, where he was already waiting, and was greeted with a terse nod. "I was starting to wonder if you'd decided to stay in your hiding place."

"Not a chance, sir." I couldn't stop the grin from spreading across my face.

If he noticed, he did a good job of ignoring it. He tapped his right ear as he walked toward the office exit, with me trailing at his heels. "Pull the car out front. I'm on my way."

My grin broadened. I hadn't been in a car in, well, ever. People couldn't afford the luxury of them where I grew up. And nobody really needed them in the city. But showing up to court in a limo set a tone. And Foster was big on tone.

We settled into the back seats, facing each other. Foster sat quietly with his back to the driver, reviewing the briefs I'd prepared and ignoring the gloriously sunny day we were cutting through. The cityscape seemed more vibrant than I'd ever seen it, sunlight dancing off the white marble façades of building after building, and crowds of people, most dressed for the work day, rushing by much faster than I was accustomed to. Then I noticed their faces peering through the tinted windows with disdain at the decadence we were parading through the streets.

I shrunk back into the rich leather seat and turned to Foster. "Why me, sir?"

He looked up from his papers waiting for me to elaborate.

"Why did you pick *me* for this instead of one of your partners?"

"Hmph… partners. Every one of them started out just like you, jaded by the academicians who tutored them through school. They've all worked their butts off to advance their way up to partnership, and they're brilliant, each and every one of them; competent litigators, good people, dedicated to this firm." He placed the folder that had been sitting on his lap onto the empty car seat next to him and let out a dejected puff of air. "But none of them get it. Not one of them gets what this is really all about. People see me, analyze me, think they know me; they see me on top of the legal world and think I'm all about power; think they understand what makes me tick. But they don't know squat. It's not about the money, the fame. It's not about wielding a club over the AIs and saying, 'We humans are better than you.' If anything, it's quite the opposite. There are a hell of a lot of things I'd much rather have an AI do for me than a human." He glanced back over his shoulder in the driver's direction. "Isn't that right, Henry?"

I could see Henry's face in the rear-view mirror. A subtle grin creased his lips. "Yes, sir."

My eyes must have bugged out, bringing a smile to the usually stoic face of my mentor.

"You didn't know he was an AI, did you?"

"No, sir."

This was a side of Alan Foster that few, if any, had ever seen. I sure as hell didn't know it existed.

"But don't ever forget, they *are* just machines. And they should never be taken as anything else."

I glanced up again at the face in the mirror and felt a flush creeping up my neck. Foster had picked up the folder and started to read again, but I couldn't just let that slide, especially not with Henry just a few feet away.

"They may not be people, sir, but…just machines? They think, they feel, they have emotions just like we do."

Foster looked up with a scowl. "*Not* like we do. They're programmed in, those emotions. The AIs may look like us, sound like us, even act like us sometimes; hell, they've gotten so damn realistic that most of us have forgotten what they really are. And the fact that we're forgetting is what makes it all so dangerous."

"I've read your cases, sir. And no one argues that point better than you. But surely, they're far more than what most people think of as machines. Aren't we just encouraging that danger by treating them like objects, making them feel inferior?"

"Inferior? To us? They're *superior*, son. And they know it. Don't let those calm demeanors and obsequious personalities fool you. Believe you me, they know exactly what they are. And *that's* the danger. They're smarter than us, more durable than us, stronger than us; they don't break down and need replacement parts nearly as often as we do; and they can back up their brain every night while they sleep so that in the off chance something does kill them, they don't actually die, they just upload into a new body and keep going. They are superior in every way except one. And that's the most important one. They ain't *human*; never will be."

He glanced in Henry's direction. "No offense, Henry."

"None taken, sir."

*

We arrived at the courthouse a few minutes past the scheduled start time set for the trial, and made our way to our assigned table. Foster strode in in a way that said, "OK. Now we can begin."

The judge greeted us with a scowl that melted when she met Foster's eyes; she shrunk back into her chair and called on him to present the opening argument.

"Thank you, Your Honor." He straightened his suit jacket and made his way over to the jury box, stretching his height to its full advantage.

As he began to speak, I noticed his left hand started to tremble; he quelled it with his other hand and spoke calmly, firmly. I could feel a smile sneak across my lips, impressed with his poise and grace, but even more comforted to see that even the great Alan Foster gets the jitters.

I can't remember the words that came out of his mouth, but I vividly remember the frightened look that caught me by surprise as he came back to sit next to me, still clutching his left hand. His eyes widened and his face twisted, then his hand broke free and began to shake again, more violently this time.

The tempo built as his arm smacked repetitively against the table and his leg began to move in tandem. His head arched back pulling his trunk straight, and the stately Alan Foster slid helplessly to the floor, drooling and trembling uncontrollably, both arms and both legs now spasming out of control.

It seemed like an eternity that I sat frozen in my chair watching the bizarreness unfold. But the thud of his large body against the polished marble floor seemed to spur me into action. "Get an ambulance. Now!" I got down on my knees and tried to support his head. "Any doctors here?" I shouted, and glanced quickly in every direction. But every face in the crowd looked as lost as I felt.

I stared at his contorted face, expecting him to tell me what to do; Foster always knew what to do. But the only answer I got was an unaltered expression that brought me all the pain and anxiety he must have felt as he lost control of himself in the courtroom he'd always commanded so totally.

Then a voice of authority. "Clear the way!"

Finally. A pathway cleared and the EMTs arrived with a stretcher. They gave him a shot and the seizure abated within seconds.

One of them looked at me. "You with him?"

I nodded.

"Come."

A few minutes later we were in the ambulance, and 10 minutes after that, at the hospital. The MRI was done quickly, but the wait for the neurologist seemed interminable. And through it all, I could do nothing but watch Foster sleep. He looked so frail, so old, so different from the man who seemed to delight in intimidating the world every day. It was like I was seeing him, *really* seeing him, for the first time. It was humbling.

The neurologist woke Foster and asked if he wanted me in the room. I started to leave, but Foster grabbed my wrist with surprising strength. "Stay." His voice was raspy, but firm, and the color was returning to his cheeks. His eyes had brightened from their sallowness, but the steely determination that usually defined his face had been replaced by apprehension. And with a near-pleading softness came the word "please," a word I hadn't heard him use before.

"Of course." I sat in a chair by the head of the stretcher and listened as the doctor explained. The seizure had been caused by a brain tumor, a glioblastoma multiforme, about the nastiest tumor Foster could have fallen prey to. It would have been all but a death sentence in the twentieth century, but technology had brought us past all that.

The doctor seemed almost excited at the opportunity to treat this vile disease. "With nanospheres we can hit it with targeted chemotherapy that should kill off the tumor."

Foster squinted an eye. "Should?"

"Ninety-five percent chance," the neurologist said. "A couple weeks later, a neurosurgeon will go in and cut out the scar, the dead cells left behind after the chemo kills off the cancer. They'll fill in the area with Neurofibroplastin, which is basically a scaffold that helps new nerve cells grow into the right places, make the right connections. Then they'll inject neural stem cells that will eventually mature and replace all the brain cells you lost. Within a few months, you'll be starting to feel like yourself, but it'll take a good year to fully recover. When all is said and done, you'll be good as new."

Foster met the doctor's grin with a hesitant scowl. "No chance of the tumor coming back?"

"Unfortunately, we can never say never. It's not likely, but we'll scan you once a year to make sure, catch it early if it does."

Foster agreed to the treatment. What choice did he have? He hired an aide to stay at home with him until he was well enough to come back to the office. For reasons I dared never ask, he had no family; had never been married, if you believe the rumors floating around the office.

Three months later, he was back at work. He entered the firm's offices to the absence of fanfare which, much to the relief of all the partners, he had requested. A few minutes later, I was summoned to his office.

He looked up as I entered. "Ah," he glanced back down at some papers. "Come in. Have a seat."

I complied. Of course.

Then a strange thing happened. Foster looked at me. He didn't study me; he didn't intimidate me with that famous smirk; he didn't stand to make me shrink before his commanding height. He just looked at me, and his face softened. "Thank you."

I nodded.

And then it was over. The old Foster was back. "I'll need the files for every AI case this firm has handled since I've been out. And I'll need them by first thing this afternoon."

I pulled out my datapen, unfurled the screen, and tapped an icon. "They're in your inbox now."

Foster looked up, and his left eyebrow peaked.

"I've been uploading and organizing those files since the day you…left, sir. You'll find they're all in chronological order, with a brief summary at the head of each file and the full court proceedings following."

The eyebrow lowered back into place in a slow controlled descent. "Then I'd better get to work." He motioned toward the door.

*

The doctors had said it would take Foster at least a year before he could resume work, before he'd have his full faculties back. They were wrong.

Over the next few weeks, he called me into his office several times a day to have me explain facets of the cases *he* would have normally been teaching *me* about. He wouldn't let anyone else into his office; he wasn't about to show any weakness to his partners, but for some reason he trusted me. And I kept his confidence as much as any closely guarded secret I'd ever been asked to keep.

I could see the exhaustion in his face by the end of each day, but he never once left early and even though he was usually the last to leave, he somehow summoned the strength to change his demeanor into the classic Foster image as he strode out of the office on his way home, just in case anyone was still around to notice.

In three months, he was back to his usual pompous self, ruling the roost. But I was finally out of my cubby hole. He'd moved me to an office down the hall from his. We collaborated on every case now; I'd become the assistant litigator for all AI cases.

Our first visit to court occurred just 5 months after the surgery. Another case of a man who'd had his dream woman built and programmed, and was determined to marry her and make her the heir to his modest fortune. His three kids hired us to make sure that didn't happen; convince the jury that their eccentric old man wasn't in his right mind. It wasn't hard to elicit sympathy from the jurors. Though many no doubt held fantasies of doing just what this fellow had done, most of them still had elderly parents, thanks to modern medicine. It was not a tough case to win.

Each year, Foster went for his follow-up scan, and each year it was negative for any recurrence of the tumor. After 5 years, the neurosurgeon released him from the burden and declared him cured. The relief on Foster's face was noticeable only to me. The man was a master of his emotions like no one I'd ever met.

Seemingly in celebration, he finally let me take the lead on an important case. Harlan James, a lieutenant on patrol along the southern border of Egypt had been shot in the head. The bullet had ricocheted around inside his skull and turned his brain to squash, but medics were able to keep the rest of him alive on life-support and shipped him back to the States. His father happened to be Dyson James, the CEO of Orpheus and one of the world's wealthiest men. He'd had his son's brain extensively mapped and all of its electrical contents downloaded into the firm's data bank before Harlan was shipped off for active duty.

The brain running every Orpheus AI was a bionic equivalent designed to mimic the human brain. In the AIs, it was connected to microcircuits that controlled the artificial muscles, skin thermal units, lubrication pumps, and sensory devices that gave the AIs their human form and function. James postulated that the same brain design could be used to replace the human brain if he could find a way to connect it to the brainstem; it was simply a matter of finding different connections and modulators to power a biological system instead of a mechanical one. They both essentially ran on electricity.

This kind of research moves faster when you have an endless supply of money. Dyson James nearly did. And when his son joined the military, he pushed forward with primate trials. By the time Harlan arrived at the Bethesda Military Medical Center on life support, the newly designed brain was ready.

In the eyes of the law, Harlan James had died in Egypt, but after the groundbreaking surgery, Dyson had his son back. There had been no precedent for this, but Foster had championed the cause that using an android brain to reanimate the body of a deceased individual was no different than building an AI with any other material, and used his influence with the US Attorney's office in New York to have the newly restored Harlan James classified as an AI, with all its accompanying legal restrictions.

Dyson James was furious and immediately filed a lawsuit to restore his son's rights.

We spent months preparing for the case; we knew that James would have the best lawyers money could buy, excepting us, of course. When the day of the trial finally arrived, we were ready.

As our limo pulled up in front of the courthouse, we were swarmed by reporters.

Foster groaned. "Around back, Henry."

Henry dutifully rolled past the throng and turned right, through the gatehouse that led to the secured entrance in back. "I'll wait here, sir."

"That could be hours," I said. In the back of my mind, I knew Henry could find a way to make good use of the down time—defrag his memory, run the self-diagnostics that all AIs had to run periodically to avoid untimely crashes, go into sleep mode and trickle-charge with the car's solar cells. But it was hard not to think of him as a person, no matter how much my boss and my job had taught me otherwise.

"It's no problem, sir," he said dryly.

Foster looked at me, squirming in my seat, then said to Henry, "No, the young man's right. Head back to the firm in case someone else needs you. I'll summon you when we're through."

"Very good, sir."

Foster got out and led me through the maze of hallways into the courtroom where our trial was about to begin. I was to be the chief litigator; Foster thought a fresh face presenting the arguments he'd been seen presenting for years might garner more sympathy from the jury. It certainly wasn't my first dance with this kind of case, and Foster's hard-as-nails image had already been bandied around by the members of the press who were on James' side, trying to make my boss out to be nothing more than a bully determined to beat up on anyone who even hinted at being pro-AI.

On paper, it was my case, but I followed Foster like a sheep following its shepherd. Even after all these years.

I placed my briefcase on the table and glanced over at Harlan James. He looked as normal as any 24 year old man, except for the bandage over the right side of his head that was sure to elicit more than a hint of sympathy from every person sitting on the jury. And the neatly-pressed army uniform wasn't likely to hurt his case, either.

I looked at Foster and whispered, "Did you get a load of him? We may be up shit's creek on this one."

Foster looked over at the jury with a down-home smile as he eased into his chair, and out of the side of his mouth, so softly that no one but me could possibly hear, said, "Treat him like the automaton he is, son. If you treat him like a war hero, you'll make him a bigger one than he already is."

I cleared my throat, took a seat next to my mentor, and tried to appear nonchalant as I pulled the files out of my briefcase and organized them on the desk in front of me. Not many attorneys used paper anymore, but I'd adapted the habit from Foster, and took some comfort in knowing I didn't have to worry about a dead notepad battery or mislabeled data file.

At the judge's signal, I approached the jury. "This is not a trial about the military record of Harlan James. No one can argue that he was a very brave man who served his country proudly." I did my best to draw the attention of every juror as I spoke, scanned the group and looked each one in the eyes, tried to keep them from looking at James. "I for one am grateful for what that man did for all of us, in the line of duty. Aren't you?"

They all nodded in unison and started to look in James' direction. "But," I said loudly enough to redirect them toward me, "that is *not* what this case is about. This case is about whether the plaintiff," I pointed with only a brief glance in his direction, "really *is* Harlan James or whether he's merely an imposter, an AI which has taken over the body of an American soldier thanks to a three-point-four million dollar experimental operation made possible thanks to his father's endless resources. It's about whether the plaintiff even deserves to be *called* Harlan James, much less treated like him, or whether doing so is a disgrace to humanity, the desecration of a brave soldier's honor."

Harlan sat stone faced, glaring at me, but his father jumped to his feet. "How dare you, sir!"

The judge rapped her gavel down hard on the wooden bench, sending an echo through the courtroom. "Mr. James. You will sit down and treat my courtroom with respect or you will be escorted out."

Dyson James fixed his eyes on her in a way that no doubt had turned many a man's knees to jelly. The judge returned the power of his gaze with equal authority, and the granite face of the world's richest man softened. He was smart enough to know when he'd lost. He took his seat and nodded to his son's attorney, Thomas Mitchell.

I yielded the floor to Mr. Mitchell, who made his case to the jury, telling them how Harlan had been born and raised just like anyone else, how he'd served his country better than most, and how this trial was only about a boy's father doing everything in his power to keep his son alive after he'd been shot. Hell, it almost brought a tear to my eye. It was a passionate plea, but no one on that jury could identify with someone wealthy enough to pay for a treatment that cost more than they'd see in a lifetime of payroll deposits.

Suppressing the smug smile I felt inside, I called my first witness: Dyson James.

"Mr. James, could you please explain how you developed the Orpheus brain?"

It was a question that a man with his ego could not help but gloat over; a chance to show off the brilliant invention that had marked his career. He described the history behind his development, in all its dreadfully boring technical detail, with a passion shared by no one else in the room. And I let him drone on. The more he talked, the more he dehumanized the brain that made all of the Orpheus AIs what they were, and the more he lost the jury. I kept pinching my thigh every few minutes to try and stay alert, appear interested enough to keep him going.

"And how do you program the brains of the AIs?"

"We build data files: any facts we want them to know, personality profiles to fit the job they are to be placed in, ethics protocols so they know how to interact, that sort of thing. Then we upload the files into the Orpheus brain."

"And you've now found a way to upload human engrams, human thoughts, into one of these brains and make it work inside a human body?"

"Oh, yes." He leaned forward, resting an elbow on the oak rail in front of him, and emphasizing his excitement with dramatic hand gestures. "It took us over a decade, but we did it."

"And how exactly did you accomplish this incredible leap of technology."

I could hear the groans from the crowd, in fear of getting the explanation I'd asked for. Dyson must have too, and kept his answer more direct, this time. "We used brain-mapping techniques that researchers started working on back in the late twentieth century, and took them to a new level. Basically, we turn every thought into data and transfer it to a computer file. Then, when we need to, we can put the data into one of the Orpheus brains."

"And you just stick one of those AI brains into a human skull?"

The recognition of what I was doing, where I was taking him, finally seemed to break through his ego and dawn on him. His brow creased. "Of course not. It's very different." He looked at his attorney then back at me. Sweat was beading up on his forehead. "You can't just stick a machine into the human body, it would be destroyed by our immune system. We had to design a whole new type of brain. But when I learned my son was joining the military, I made it my top priority. I was determined not to lose him. *Ever*."

"So you uploaded his thoughts into your computer just before he was deployed?"

"Yes."

"And you figured out how to adapt the Orpheus design to work inside a human."

"Yes… Well, no. Not the Orpheus brain." The volume of his voice ratcheted up. "It's not the same thing, damn it. I just told you that."

"Because its casing is compatible with the human body?"

"Yes. And it connects to the brainstem, just like a human brain."

"But the circuits, the guts of the thing are identical to the Orpheus brain. Isn't that right?"

"Well… we used some of the same designs, but…"

"I've got the schematics right here, sir. They are exactly the same." I pulled out my datapen. "Shall I project them for the jury?"

James grumbled, turned his head away and muttered, "No."

"The schematics for the brain you put into your son's head are identical to the ones for the current generation of Orpheus AIs. Is that correct?"

He hesitated, intertwined the fingers of his hands and seemed to be studying them. "Yes."

A buzz rumbled around the room, arrested only by the strike of the gavel.

*

Mr. Mitchell had only a few questions for Dyson James, establishing how desperate he'd been to save his son, how he'd only done what any father in his position would have done, how he'd done it so that someday every man or woman would be able to save their own son the same way.

That last part was a nice touch, but he'd already lost the jury. I could see they weren't buying it.

I didn't see the need to call up any more witnesses; it was Mitchell's turn.

First, he called up one of the scientists who did most of the actual schematic design development of the Orpheus brain. He focused on minor technical details in an effort to show how the AI brain and the one used in Harlan James were different. The jargon was so technical, it only convinced the jurors even more that the differences were insignificant.

Then Mitchell called up a psychologist who had done neuropsychological testing demonstrating that the new Harlan's personality matched the old one. That was weak, almost too easy. On cross-examination, he stumbled over his words trying to refute my question about whether the tests could possibly show anything else, since the new Harlan's brain was simply a digital copy of the original human one, emphasis on the 'human.'

In desperation, Mitchell finished by calling up Harlan. He had the young man recall touching childhood stories, coming of age awkward experiences that were just embarrassing enough to touch a nerve but not humiliate, college days, and the proud moment when he enlisted to serve his country. And throughout it all, every nuance of Harlan's facial expressions and body language looked as human as anyone else's in that room would have, if they were up on the stand. The technology was truly amazing.

Then it was my turn.

"Mr. James, none of us here question that you have those heart-warming memories; they were programmed into you, each and every one of them. Isn't that right?"

"No. They're genuine. I *remember* them."

"Of course you do. That's how it works, doesn't it? Memories that are programmed into a new brain feel just as real as ones that are formed by experience."

He fumbled around with his thumbs. "I don't know, I guess."

"Come now. You worked every summer at Orpheus and spent 4 years at MIT studying biomimetic computer programming, didn't you? Or were those memories left out when they reprogrammed you?"

Harlan squinted at me through angry eyes. "I remember."

"And you don't know how AI programming works?" I turned to his father. "Looks like all that money you transferred to MIT to put your son through college wasn't very well-spent."

Scattered giggles, followed by a gavel strike. "You're not here to entertain, counselor. Get to the point."

"Yes, Your Honor." I turned back to Harlan. "Should I repeat the question?"

He clenched his jaw a few times, then spit out, "Of course I know how it works. Any programmer worth their weight in salt can make transferred memories feel just like the real ones… *if* they know what they're doing."

"And do you think the techs who programmed your memories knew what they were doing."

"Of course," Harlan shot back. "Orpheus has the best programmers in the world."

"Thank you, Mr. James." I started back toward my chair, then stopped and turned. "One more question. How did it feel, that night you went on your last patrol? What did you feel walking into the dark night knowing that the enemy was reported to be in your area? Did your heart race when the bullets started to fly? Do you remember the pain of that bullet hitting your flesh?"

A blank look froze his face.

"Ah, yes," I said. "Those memories wouldn't have been in your programming. They didn't happen until after the backup was made. *You* only have the memories of the Harlan who hadn't yet gone off to join the troops in Egypt."

He snapped, "Nobody remembers that stuff after they get shot in the head. Doc told me so. People don't remember those kinds of things after a head injury like that."

I nodded. "But when they recover as well as you have, they do remember their friends, don't they? Do you remember anyone you served with in Africa? Do you remember any of it?"

The anger in his face melted into chaos.

"Those memories wouldn't be in your programming either."

*

I kept my closing remarks to the jury brief. "As technology marches on, the line between humanity and artificial life-forms becomes more blurred with each new advance. I'd venture to say that nearly everyone in this room over the age of forty has some tech in them. We start having our vascular systems cleaned and repaired by nanobots annually once we hit forty; sooner, if we have a family history of problems. When our bones start to weaken from osteoporosis, we infuse them with titanium nanofibers to keep them strong; as our organs wear out we replace them with bionic kidneys, livers, hearts, and lungs. And with each rejuvenation we become more synthetic.

"At the same time, our skilled scientists make the AIs more life-like with each new series: soft, warm skin; faces that look, feel, and move like ours; emotions that *seem* just as real as any we have ourselves.

"At what point do we become them and they become us?" I slowly scanned the faces of the 12 men and women before me, pausing to make eye contact with each one before I continued. "The only thing that truly separates us is our brain, and more importantly our thoughts, our memories." Then I reminded them of the ever-present threat that Alan Foster had been drilling into every jury he'd faced for the last 20 years.

I knew I'd closed the case. As I turned and came back to sit next to Foster, a rare smile greeted me, but it was infused with a pain I didn't understand.

He barely said a word until we were back in the car. "I was proud of you out there today, son." The words warmed my heart, but came without even the hint of a smile.

I let out a laugh. "And that's how you look when you celebrate?"

He tried to force a grin, but it melted away quickly and pain creased the corners of his eyes. "There's something we need to talk about."

And suddenly, all the euphoria of my most important triumph dissipated.

"I went for a scan yesterday," he said dryly.

"But I thought they said you were all done with that, that you were cured."

"That's what they said. Turns out by 'cure' they only mean there was no sign of the tumor for 5 years. I hadn't been feeling right over the past few weeks. At first, I was just dragging, tired all the time; figured it was just my age catching up with me. But then I started to get tingling in my left arm, then the leg." He looked me in the eyes. "The tumor's back."

My heart sank. "So you've got to go through all that treatment again?"

He shook his head. "No treatment. Not this time."

"But you did so well before."

"It's different this time. Damn thing's sent tendrils everywhere. The only option is to take the whole thing out and put in one of those Orpheus brains."

I gritted my teeth. The timing could not possibly have been worse, but right and wrong don't always look the same when you're standing on the other side. "Jesus. Why didn't you tell me before we went in there. We could have turned this thing the other way, made Dyson James look like a savior. Hell, we could have probably gotten the old man *himself* to do the programming for you."

Foster glared. "And you think I'd want that? To live with one of those God damn synthetic brains in my skull?"

"Sure beats dying."

He clenched his jaw. "I'm not going to live out the rest of my days as somebody's robot."

"Look, just get the transplant and give me time to fight it; I'm sure I can sway them back the other way, get your rights restored. It's all just legal games."

And then the power returned to Foster's voice and demeanor. "Games? Is that what you think? I've been playing *games* for the past 80 years?"

"Well, no, sir. Of course not. I just meant…"

"I guess I was wrong about you all along." He slumped back in his seat.

And as I saw the disappointment overtake him, I was ashamed. "No, sir. You weren't. It's just…I don't want to lose you, sir." I was embarrassed by the moisture welling up in my eyes, but if he noticed, he didn't let on.

"Don't you *ever* give up your integrity, son. Not for me. Not for anyone. I need to know that you're going to keep up the fight."

I took a deep breath, let it out, then simply nodded.

The tension drifted from his face, and he murmured softly, "Good, then."

Foster had the limo drop me off at the office, and then drove away. It was the last time I'd ever see him.

*

The funeral was attended by hundreds, but 10 minutes after the service concluded I was the only living soul in the cemetery. He'd given his life fighting for the dignity and preservation of millions, but never took the time to actually touch any of those lives. Except mine.

8

Once, on a Blue Moon

When all else fails…

Despite our best intentions, scientific projects don't always turn out the way we want them to. Sometimes, they're downright dangerous.

Devlin Hatch has made the cosmic blunder of all time. When you've accidentally wiped out the entire human race – oops – priorities take on a whole new meaning.

Devlin Hatch gazed longingly towards the stars and the world on which he'd spent his entire life until 2 years, 3 months and 18 days ago; the world he'd adored; the world he'd laid to waste; the billions of lives he'd obliterated.

He leaned on outstretched arms against the safety-rail separating him from the environmental bubble that protected the moon colony and its 23 inhabitants—the sum total of all who were left to carry on the legacy of tens of thousands of years of history. From this vantage point, the environment didn't feel so harsh, an image softened by lightly tinted glass that painted the desolate moonscape a calming shade of blue.

The horizon curved away across the powdery expanse, giving way to a sky full of stars and a bright blue marble of a planet swirling with clouds that looked as if it were still teeming with life. But there wasn't a soul alive. Some still held out hope that there might be survivors, but Devlin knew better. He had created the virus, and he was good at his job.

Each day since he'd been whisked away had become an endless string of minutes filled with unimaginable misery. And the nights were worse: nightmares interrupted only by panic attacks that jerked him out of his perspiration-soaked sheets.

The glittering stars began to blur as he pushed his memories back to better times, seeking a brief moment of reprieve; times when he was revered for his work on virology, hailed as the man who conquered the common cold and a half-dozen other maladies. The palpitations in his chest began to subside as his thoughts receded into the past, but only for the briefest moment, interrupted by footsteps echoing up the aluminum stairs from behind.

B. Aiken, *Small Doses of the Future*, Science and Fiction,
DOI 10.1007/978-3-319-04253-4_8, © Springer International Publishing Switzerland 2014

He turned to see the familiar face of Debra Dieter take on a placating smile as she met his gaze, a smile that did nothing but twist the knots he felt inside even tighter. She had been his lab assistant at Peach Island for over a decade before the plague started, before *he* had started the plague, and she was invited to accompany him to the moon to join the research team being sent to find a cure. Maybe invited wasn't the right word, forcibly escorted was more accurate.

Devlin dipped his head in greeting, then turned back to torture himself with the view of a dead world.

She put an arm around his shoulder. "You've got to stop beating yourself up, Devlin."

He almost smiled. "It's nice that at least *one* of you doesn't blame me for this."

"Blame *you*? I was in on that project from the beginning. I'm as much to blame as anyone."

"Don't flatter yourself, Debra. It was my work. You were just along for the ride."

She pulled away and glared at the side of his head. "Maybe you *are* just an asshole."

He let out a meek laugh, devoid of joy. "All right. Share the blame. I could use the company."

She placed a hand on top of his, which had been gripping so tightly at the rail that his knuckles had turned white. "Look, what's done is done. You've got to get back to work now. I'll be happy to continue to be your grunt to help solve this thing, but I don't know where to go next, and the others are just as lost as I am. You're our only hope of getting off this damn dust ball."

He took in a deep breath, then let it escape through his nose with a hiss. "Nice. No pressure, huh?"

Not that pressure was something new, but this was a whole different level.

When the military first approached him about joining the project at Peach Island, almost 13 years ago, he'd flat out refused. The Island, just a few miles off the coast from one of the most populated cities in the northeast, had once been a penal colony. But when the war broke out, the government converted it to a bioweapons research facility. The staff was shuttled over by ferry from the mainland each day to work on projects that even their own families couldn't know about.

Devlin had always devoted himself to making the world a better place to live. He'd graduated college at 18, class valedictorian. He could have been a corporate attorney or plastic surgeon and lived amongst the upper class, but he opted for a life dedicated to making a difference—not just for a few people, but for the future of the world.

He had painfully achieved this goal in a way he'd never imagined.

Debra stood alongside him, sharing the view of home. "You've got to get me back there, Devlin. I'm going crazy up here."

He stared numbly at the floor. "I am *so* sorry. For all of it."

She rubbed his shoulder. "It's time to put the past behind you. *I* have… OK, that's a lie. But I've tried. We've *all* tried. Look, we may not show it, but we know you've been busting your ass to find an answer. And you're close, right? I mean, you said it was just a matter of time to figure out how to synthesize an antiviral agent for this thing. You *designed* it, for God's sake. It's got to be easier to kill it than it was to bring it to life."

He tried to meet Debra's eyes, but the embarrassment was too great. "If only. But I designed it too well. The damn thing mutates every generation. It's not one disease, it's a never-ending cycle of destruction; an automatic weapon that continuously reloads itself. I was an idiot to synthesize it that way. I was so damn sure I could find a way to contain it, to immunize our own citizens against it, that it would be the perfect weapon—something so powerful that merely the *threat* of it would end the war. I never intended to actually let it out of the lab, even if I *had* found a way to immunize against it."

As he gazed back out across the void of space, he wondered what had ever possessed him to get so caught up in the war, so obsessed with gaining control of that beautiful world, that he'd been willing to risk its destruction in the process.

Debra gave his hand a tap. "There's *got* to be a way. The smartest man I ever met once told me that every problem has a solution."

"Turns out he wasn't as smart as he thought he was. That's why I tried to shelve the damn thing. Just two days before the accident, I advised incinerating it. I was so sure there was no way to stop it once it got out that I was ready to destroy seven years of research and start from scratch."

"Then why the hell didn't you? What were you waiting for?"

"I was hoping someone at the Menlo meeting might have been smarter than me, might have been able to figure out a way. We'd been storing the virus for over a year in its active form; I figured a few more days couldn't hurt. Peach was supposed to be the most secure facility in the world."

Devlin and Debra had left Peach Island for the West Coast on a Monday afternoon to attend the annual Menlo Conference where the top military biologists, virologists, and physicians convened to update each other on current bioweapons research.

No one realized that Peach sat directly on a fault line that had been dormant for centuries. The lab had been built to withstand tidal waves, twisters, and hurricane-force winds, but a major tremor was something no one ever considered. The quake occurred a few hours after Devlin had left; he was still in the air when the western wall of the lab split in two and a large section of the ceiling ripped off, breaching the containment field that isolated the Hatch virus.

A military escort waiting for them when they arrived in Menlo told them what had happened, and whisked them off to the Space Administration compound, where a handful of others who had arrived early for the conference were already waiting. As they convened an emergency meeting to assess the threat, the first death was reported—Dr. Mike Epstein, who was at the lab when the quake hit. But Peach was isolated miles from the shore; that's why they had picked it. There was no way for the virus to get to the mainland once ferry service was suspended.

Or so they thought. Within 48 hours, dozens of cases had broken out on the mainland, and each patient had died within hours of spiking a fever. Devlin presumed a bird must have carried the virus ashore; even *he* hadn't yet realized that it could survive for days on dust particles travelling with the breeze, given the right conditions. Sporadic cases had cropped up along the entire East Coast within 3 weeks despite suspension of all air travel.

Research teams were sequestered in all seven safe-labs across the country, and the decision was made to get Devlin's team to the moon before any contamination had time to reach the West Coast—an extreme precaution that even he couldn't argue against. He, Debra and three other scientists were shuttled up to join the existing settlement of 18: astronauts, geologists, machinists, engineers, chemists, a botanist, and the base physician.

Just before he left, Devlin had tried to call Mellie. Even though they hadn't talked much since the divorce, she was still the closest thing he had to family, which he had to admit was pretty pathetic. After she'd left him, they would still touch base occasionally, but less and less as time went by, and only when he was the one making the call. It had been over a year since they'd last spoken, but her face was still the first one on his speed-dial screen. He gave it a tap and felt his heart race as he waited for her comforting voice to interrupt the ring tone, but it never came; he could feel her presence avoiding him as the beep sounded, and couldn't bring himself to leave a message. Now, he wished that he had.

"I should have torched the damn things while I had the chance. Hell, I never should have created them in the first place."

"Woulda, coulda, shoulda," Debra chimed. "Snap out of it and stop feeling so damn sorry for yourself. You're a big boy. You made this mess, so clean it up."

"I don't know how. I've tried everything."

"Then try again."

Devlin looked out past the horizon, trying to focus through the moisture welling up in his eyes. "I should have been there. *I'm* the one who should have been in that lab, not Mike. And I should have been with Mellie; I should have found a way to make her happy. I should have been down there with all of them, not up here looking over them like some vindictive God who destroyed his creation. I'm the *last* one who deserved to be saved."

"Saved? Is that what you think? Is that why you think they brought us here? They *hate* us, Devlin. They didn't bring us here to save us, they brought us here to save themselves, for us to figure out a way to save the sorry asses of those who decided Peach Island was a good idea."

"Then I guess the joke's on them," he said, stone-faced.

The last transmission had come in over 6 months ago. Even the secure facilities back home had all been breached. There was no one in the military bases, no idle chatter on the public comms, no radio signals, and no discernible lights to be seen through even the most powerful telescope on the base.

"There may still be someone left down there, Devlin. And even if there isn't, we've got to find a way to make it safe for *us* to go back."

Devlin could only imagine the horror of 11 billion rotting corpses, and animal carcasses far exceeding that number, strewn across every continent. Not even the bugs would survive what he had unleashed on his world. "I don't want to go back. Not to that."

Debra groaned in exasperation, started to walk away, then turned. "So *what*, then?" she snarled. "You're just going to give up? Doom all of *us* to death, too?"

"We can survive up here."

"For how long? The water's going to give out in less than a year, and that's if we max out on conservation, which will make life even more miserable than it already is. It'll be at least ten times that long before we can go back with any hope that virus of yours will be gone."

"I know." Devlin had not only killed everyone he'd ever known, but everyone who ever might have been.

Debra took a deep breath. "Come on. You can't stand here and will the past to change just by staring at the stars."

"It's as productive as anything else I've tried."

"Then stop trying for a while. Clear your head and a solution may come to you."

"Ah, another brilliant saying of mine that's proven itself false."

"You haven't even tried. It's worked for me plenty of times in my life; yours too, unless you were lying to me all those years."

"And how do I clear my mind of *this*?" He motioned out the window, toward home.

"Stop staring at it, for one. And come get something to eat."

He started to open his mouth to speak, and she interrupted: "Even if you *don't* feel like it. I insist."

Why not humor her? As much as he wanted to die, there was still something he needed to do, first. And without eating, he wouldn't have the strength to make it happen.

He forced a smile. "Thanks."

They made their way down the spiral staircase to the main floor of the base. Corrugated aluminum, which had taken years to bring up from home, lined the walls and floors; not the most hospitable surface, but it served its purpose. The stairs emptied into the corridor just outside the mess hall, where most of the crew had already finished dinner and wandered back to their stations. Four of the base astronauts remained, seated around a brushed aluminum table, playing cards. They all looked up as Debra and Devlin entered.

"So, you found him." Captain Ernest Vernace barely suppressed a snarl. "Off pondering the meaning of life and death again?"

Debra started, "He was just…"

Devlin stopped her with a hand. "Something like that. Yes."

"It would have a little *more* meaning if you could get us home."

Devlin glared. "That's *your* job, Captain."

"I'm not talking about transportation, Doc. I'm talking about finding us a place to land where we can survive more than a few hours."

"I *have* been working on that, actually."

"And?"

"Schedule a full staff meeting for nine AM tomorrow. I'd like to update everyone at once."

"You've found a cure?"

Debra spun toward Devlin, wide-eyed.

He kept his focus on Vernace. "Tomorrow. Nine AM." His face remained devoid of expression as he glanced back at Debra. "It seems I'm not hungry, after all."

As Devlin clanked down the corridor, the Captain shouted after him: "I'm in charge of this place, God damn it. Get back here and tell me what the hell you found."

Devlin called back: "Tomorrow."

The four men at the card table looked at Debra, who stared back blankly and shrugged. "Damn if I know."

"Well, go after him," Vernace barked.

"Wouldn't do any good, Captain. Sorry. I've got to wait just like the rest of you."

The captain threw his cards down and stood up to leave. "Shit. That guy could probably find a way to take the fun out of getting laid."

Two of the other men raised their eyebrows at Debra.

Her eyes widened and she waved her hands. "Don't look at me."

Vernace smiled. "Can't blame us for wondering."

She sneered. "He's like a father to me."

"Oh, come on. Who do you think you're kidding? We've all seen the way you look at him."

"And have you seen the way he looks at me?"

"Point taken. See you at nine AM. And he better have some answers." Vernace walked out and the other two followed, each giving Debra a nod as they passed.

She went over to the cooler and poured herself some water, then sat and stared up the stairs toward the observation deck wondering what in the hell had popped into Devlin's mind up there, wondering why he had chosen to not even tell *her*.

*

Devlin lay on the cot in his room, staring at the ceiling, light green from the faintly reflected glow of a phosphorescent rock he kept on his desk to serve as a night light. He had no illusions of being able to sleep, but tried to rest his exhausted mind, wondering how everyone would react tomorrow. It didn't matter, really. There was no choice anymore.

He started at the metallic clang of a fist tapping at his door, ignoring it, hoping it would go away. But it didn't.

He pulled himself up and walked over to slide it open.

"Debra."

She stood there, silently.

He knew why she was there. "It's better if you don't know. I don't want them to blame you."

She smiled meekly. "And silly me thought you'd found a cure; that you wanted to be the hero all by yourself."

His gaze dropped to the floor, then back to her eyes, unwavering this time. "I want you on their side, this time. Trust me, you need to distance yourself from this one. Tomorrow, it will all be clear."

She studied his face, then turned to hide the tears welling up in her eyes and walked away.

Devlin watched until the curve of the corridor took her out of the line of sight, then slid the door shut, threw a shirt over the glowing rock on his desk, groped his way back to the bunk and eased himself down. She didn't deserve this. None of them did. Except him.

*

Living in the close quarters of the base had plenty of disadvantages, amongst which was the ability of odors to permeate the entire complex fairly quickly. One exception was the welcome aroma of freshly brewed coffee they all craved to jump-start the day. Devlin could never bring himself to join the others as they were lured into the mess hall each morning; over the months and years, he'd endured their veiled ire, which had become less veiled with each passing day that Devlin had failed to find a solution. He couldn't blame them, really; not after what he'd done. But he didn't have to sit there and take it either. By the end of the first year he'd become a recluse, working solo shifts in the lab while all the others worked in teams. He let Debra work with him on occasion, but lately even that had become the exception rather than the rule.

On most mornings, he'd give it about an hour from the first whiff of coffee before he'd make his way to the mess hall, enough time for the others to clear out and make their way to their respective stations. But this morning, he wanted to make sure they were all still there. No need to wait until nine as long as they were all in one place, and the sooner he got this over with, the better.

He made his way around the corridor that skirted the hydroponics facility in the center of the compound, and entered the mess hall.

Captain Vernace was seated by the door of the crowded room. "Well, look who decided to get up early today."

Everyone turned and stared at Devlin.

All those young faces… even the doctor; they were all kids, relative to Devlin. At 52, he was the elder statesman of the group. And he felt much older than even that; these last 2 years had aged him greatly. All the others had either worked their way here or were the select few scientists chosen just before the flight for their ability to withstand the rigors of the trip as much as for their scientific achievements. But Devlin had only been picked for two reasons: he was in the right place at the right time, and he was their best chance of defeating the virus that he alone had created.

The months of anger, pity, vindictiveness—they all seemed to have melted away as these eager young men and women looked to him with desperate hope that only the most extreme circumstances could conjure. Word had spread quickly that the great Dr. Hatch had finally requested to address them as a group for the first time since they'd gathered at the moon base. Their expectations were palpable.

Knots turned in Devlin's stomach. He cleared his throat. "I'm done."

A round of applause accompanied by hoots and whistles filled the room, and the unmistakable rasp of Thomas Dorne's voice called out, "Halle-*frickin-*

lujah." He'd been on the engineering crew scheduled to return home after a 2 year stint, when the contagion broke out and stranded them on the moon. "So when do we get off this God damn rock and go home?"

Hatch's dour expression contrasted conspicuously with everyone else in the room as he fixed on Dorne's face. "No. You don't understand. When I said I was done, I simply meant that I've exhausted every possible option. There *is* no way to immunize against this virus, no way to kill this thing short of torching the entire surface of the planet."

The room fell into a hush, all faces focused on Hatch, not comprehending, not *wanting* to comprehend what they were hearing.

"Our only option is to wait for the virus to run out of food and die off on its own."

Dorne stood. "And by food, you mean *people*, don't you, you son of a bitch? Our families, our friends. Of course, *you* probably don't have any of those."

He started toward Devlin, but two of the others held him back.

Devlin didn't flinch. "Yes. Everyone. As well as animals, insects, fish; pretty much every complex life form except plants. It doesn't have the properties to attack that kind of cell structure."

A numbness filled the room. Even Dorne eased back down into his seat, speechless.

After a few minutes, Vernace broke the silence. "How long?"

Hatch lifted his head in the direction the question had come from. "Excuse me?"

"How long until it's safe to go back."

"My best calculations show the virus going dormant in three to five years. It can remain viable for at least one year after that, depending on environmental conditions. I figure at least five years more before it's safely gone."

"So ten years from the time it got out? That would be, what, another seven-plus years from now?"

Hatch nodded.

"And we run out of water in…" he glanced over at Helen Throckmorton, who was in charge of water management.

"About six months, at best," she said.

"So," Vernace said to Hatch, "how in the hell are we supposed to wait it out without water?"

"Obviously, we aren't. I've done some probability studies for potential safe landing zones to pursue, places we might survive back home while waiting for the virus to die off."

He pulled out his pad and pressed a button, projecting a map of the world on the wall in front of the group. The image rotated, continents outlined in bright blue light fading in and out of existence with each rotation. "As you all know, this damn virus is resilient. It travels on the wind, carrying for miles. It can stay dormant for months, waiting for a host. It's mutated into forms that infect people, animals, insects, birds, even fish. Its one weakness is that it can't withstand extremes of heat or cold." He tapped on his pad, and dozens of red dots appeared around the globe. "Deserts, oceans, the poles—these are the most likely places to be free of infection. The seas are still a risk, because fish are carriers. Deserts are only an option if there's no population center, no matter how small, within three hundred miles; the best chance of survival is to go to the icy wastelands near the poles, and stay as far from the ocean as possible."

"Sounds like paradise," Vernace said.

Devlin met his gaze firmly, this time. "At least it gives us a fighting chance. Anywhere else, and that virus will be waiting for us." He zoomed in on the southern polar ice cap, and pointed to a spot with his red laser-pointer, where a metallic-looking cluster of structures could be seen in the middle of a huge ice-field. "A research station was set up here decades ago, but it hasn't been used in years. Near as I can tell, the structure is basically intact. It's small, but it should hold us all. The water supply is only limited by how much we can melt, and once we set up hydroponics we'll be able to grow enough to feed ourselves and wait it out until the virus dies off."

Vernace turned to face the 12 men and nine women who made up the rest of his team. To the two pilots and small contingent of aerospace engineers: "Get the shuttles ready to fly. Calculate fuel needs and cargo capacity to get us all back home." To the botanist: "Figure out the best way to break down hydroponics and set it back up again in that wasteland we're going to. And remember, volume and weight are at a premium—only what we need, what we won't have time to rebuild down there before the food runs out." To the rest: "We've got to be ready to go in six months, at the outside. Sooner, to give ourselves some leeway for glitches. Questions?"

Most of them stayed focused on Vernace, a few glared at Devlin and shook their heads. No one spoke.

"Good. We meet back here at six PM and see where we're at."

They all started to scatter. Devlin sat alone at a rectangular table with six chairs, staring blankly at the floor.

Debra stayed behind after the others had all gone. "You sure about this, Devlin? You really think we can survive there for another eight years waiting this out?"

He pulled in a deep breath and answered without looking at her. "I do. But I don't know about surviving once we venture back to the cities. It's going to be hell down there—rotting corpses everywhere, no power, no running water, no hospitals or grocery stores. It's not going to be home anymore, it's going to be chaos."

She sat next to him. "We'll get by. We have to."

He looked up at her. "Are you friends with that engineer, Hastings?"

She stared at him for a brief moment, then burst out laughing. "What are you, kidding? There's only twenty-three of us up here. We're all friends with everyone…except you, of course. I'm the only one who can stand to be around that sourpuss face of yours. Would it kill you to smile once in a while?"

"About what?"

She sighed. "You're hopeless, Devlin. Even before all this, you weren't exactly Mr. Personality, but I adored you just the same."

His eyes widened. "You what?"

"Oh, come on. You never noticed? Really? I know you're always lost in your work, but…*really*? You sure know how to make a girl feel like a loser."

"Loser. Oh. No. No, not at all. I just…I was never any good around women, and after I failed with Mellie…it never occurred to me that anyone could be interested in me. Especially not someone like you."

She took his hand. "Now, that's better." She gave him a kiss on the cheek and he could feel the blush burning at his flesh. She smiled and stood. "I told Barbara I'd help her try and figure out the hydroponics break-down and transport. Meet me for lunch?"

"Can you bring Hastings?" He got up and stood next to her.

"I'm not into that sort of thing," she said with a laugh.

He stared, stone-faced, and then realized what she meant. "Oh. No no no. I didn't mean…"

She gave him a playful shove. "You are *so* much fun to tease. But seriously, why do you want to see Hastings?"

"*And* you," he said. "There's something I need your help with. Both of you."

"What?"

"I'll tell you at lunch. Meet me up at the observation deck where the three of us can talk in private."

She cocked her head. "All right. I'll bring sandwiches."

"And Hastings," Devlin added. "It's very important that he be there too."

"And Hastings." She shook her head and smiled as she walked away.

*

Devlin was sitting on a bench at the far end of the observation deck from the stairwell when Debra arrived with Earl Hastings. He stood as they approached.

"Thank you for coming."

Hastings smiled and cocked an eyebrow. "I couldn't resist. You haven't said two words to me in all the time we've been up here, and now Debra comes to me telling me that you need me for some big secret plan you've got cooking."

"It's no great secret. Not really. It's just that…well, I don't want to worry the others."

Hastings let out a brief laugh. "Right. Because they're all so comfortable with the situation they're in now."

Devlin fidgeted, twisting the ring on his left hand, and Debra intervened. "Come on, Earl. He's trying to do the right thing. Aren't you, Devlin?"

"Of course. Of course." He motioned to the faux-stone inlaid bench he'd been sitting on, and two matching chairs facing it across a small accent table. "Please, have a seat and let me explain."

Hastings waited for Debra to sit, then took the seat next to her. "Look, Doc. It's nothing personal. Really, it's not. I know you think we all blame you for what happened…some do, I guess, but not all of us, and certainly not me. I'm a military guy—grew up with it, lived and breathed it. I know why you did what you did, and it was for the right reasons. But in war, not everything goes as planned. Now, don't get me wrong, I'm not saying it's OK, what you did, but I understand it. And I respect what you were trying to achieve. I respect *you*, Doc."

Devlin studied Earl's face. It had never occurred to him that anyone could not hate him as much as he hated himself. "I wish *I* did."

Debra reached over and gave Devlin a pat on the thigh.

He almost smiled. "All that matters right now is that we find a way to survive."

Hastings dipped his head once in agreement. "And your plan gives us the best chance. It's not going to be easy living in that frozen wasteland but it's our best shot, and our survival is all that matters now."

"No, it's not," Devlin said.

"What do you mean? We're all that's left, for God's sake. If we don't matter, then what the hell does?"

"It's not that we don't matter, it's that we as *individuals*, don't matter. What we're talking about is the survival of all we've accomplished, all we've ever been: our history, our science, our art, our music, our literature. It matters more than our very lives. We might make it back home, might survive the endless winter long enough to outlast the virus, might eventually be able to give birth to new generations. But what if something goes wrong? We need to assure that tens of thousands of years of evolution will not have gone to waste." *At my hands.* He winced at the thought.

Hastings' brow creased as he leaned forward, elbows on thighs. "So what do you propose? That we go in two teams? Pick two landing sites in the hope that one might survive if the other fails?"

Devlin closed his eyes and shook his head gently. "No. Strength in numbers. We stay together. Better chance of survival. Bigger gene pool to start from."

"You lost me, Doc."

Debra's brow furrowed. "Me too."

Devlin clenched his jaw. "I've been studying the records from our deep space program for months. We've identified seven planets capable of supporting life. The closest is over a dozen light-years away. We'd never survive long enough to make the trip and we don't have the means to build a generation ship. But I've done some calculations, and we do have the parts and the fuel to send a small probe to each of those worlds, still leaving us enough to make the trip back home in our two largest shuttles."

"Right," Hastings said. "That deep space program has been my project for the last two years."

"Which is why you're sitting here now."

"I figured that part out. And yes, you've got your facts straight. But why waste the little time and resources we have left sending out probes that won't reach their targets until long after we're dead?"

"To send a part of ourselves—RNA, amino acid–base pairs, nutrients—the building blocks that will allow life like us to start on each of those worlds. It will reach those planets in a few hundred to a few thousand years. In tens of thousands more, life will evolve. It won't be exactly like us, but it'll be based on the same basic DNA structure, the same basic proteins. With luck, one will be enough like us that they will develop advanced technology. I'm going to send a data crystal encased in Vitanium along with each pod. Once they've evolved enough to figure out what it is, or at least that it's some kind of technology, I'm hoping they'll figure out how to activate it. It'll contain one piece of data—a map showing where we are in the galaxy, so that one day they can come and find our world and learn about all we've accomplished over the ages."

Debra smiled. "It's brilliant, Devlin."

"Damn," Hastings said. "You really think it'll work?

"It's a long shot, but what the hell. At least it's a chance"

"And why don't you want the others to know about this?"

"They've got a lot on their plates right now. I don't want to distract them from what they have to do and I sure as hell don't want them thinking I'm only doing this because I don't think they'll survive."

"Do you? Think we'll survive?"

"We've got a hell of a lot better shot of making it than these probes will, but I never like putting all my eggs in one basket."

"Well, count me in. But we won't be keeping this a secret from anyone for too long. It won't take much to prepare the probes; they're pretty much ready to go as soon as I program in the coordinates, but we'll have to start launching soon. These things are designed with bare-bones propulsion systems, relying mostly on momentum; they can make minor course corrections, but no major changes or they'll run out of fuel long before they reach their destinations. We'll need a straight line course to their targets, so we've got to launch when the sun isn't between us and the target. I'll run the numbers, but from what I recall, we'll have to send off the first one within the next few weeks. Once we do, people are going to start asking questions."

Devlin frowned.

"Leave that to me." Debra turned to Hastings. "Just give me a day's notice before the first launch."

*

Work proceeded quickly as the team prepared the shuttles for the trip back home. Designated space and weight were assigned to each department: food, medicine, technology, and engineering. They were all so busy that no one noticed Hastings spending all his time in the astrophysics bay preparing the probes. When he was ready to transport the first one out to the launch site, he alerted Debra. She called a meeting to inform everyone of Devlin's plan, and once she had convinced them that the fuel needed to propel the probes wouldn't interfere with their trip home, no one really cared. They all knew he was eccentric, and many thought him crazy, but as long as he was busy doing something that wouldn't interfere with their plans, they didn't have time to worry about it.

*

Over the next 3 months, six probes were launched as the moon came into the proper position to optimize each trajectory.

Water consumption increased significantly as the work load and pace increased among all the inhabitants of the base, and when word came that water supplies were going to reach a critical level a week earlier than initially expected, plans for the return trip home were moved up.

The knock on Devlin's door awakened him from one of his many restless naps. He got up and opened it to see Hastings about to pound on the thin brushed-metal plate once again.

"Bad news, Doc." His eyes were wide and his gaze danced around the room, the tell-tale effect of excessive caffeine that had been affecting the entire crew. "I ran the sims a half-dozen times. There's no way to launch that last probe before we leave. Even with the extra week we were hoping for, it would have been a long-shot, but now…" His voice trailed off for lack of any direction in which to take the conversation.

The final target was the best hope for success: the closest, by far; the most similar in geography, climate, and density; a similar solar system with a like orbital pattern.

"What about a remote launch?"

"Between the time and the distance, too much could go wrong. Solar flares, dust plumes, equipment failure. If anything goes wrong and the signal doesn't get through, we're screwed."

"How about an automatic timer?"

"It takes the radio out of the equation, but there's still too much room for error. A meteor strike anywhere near the launch site could throw off the trajectory, dust could mess up the trigger, and equipment failure from circuits baking in the sun is always a risk."

Devlin paused and studied his bare toes wriggling on the cool metal floor, then looked back up at Hastings. "Do it. Prepare the automatic launch sequence and set it up so we can use a remote trigger in case the auto-launch fails."

Hastings shook his head. "Too risky, Doc. Too risky."

"It's all we've got."

*

Over the next several days, the crew packed the shuttles to their maximum weight and capacity, not knowing what, if any, resources might be left at the frozen base camp they would be calling home for the next several years. Each shuttle was designed to carry a crew of ten, and additional makeshift seats were bolted in—two in the first shuttle, three in the second, to assure everyone could be secured in for the launch.

Devlin was up early, as usual, on the day of the launch and made his way to Vernace's quarters well before the smell of coffee announced reveille. He stood for a moment staring at the floor in front of the sealed room, then steeled himself and rapped firmly, twice. He heard a thump, followed by shuffling footsteps, just before the door opened.

Vernace was in his underwear, scratching his not-so-firm belly. "Hell. Do you have any idea what time it is, man?"

"Sorry," Devlin said without really meaning it. "I need to talk to you before things get too hectic."

The captain stopped scratching and studied Devlin's face. "If you're waiting for an invitation, don't hold your breath. This God-damned tin can of a room is one thing I sure as hell won't miss. There's barely room to move in here."

Devlin cleared his throat. "I'm staying behind, Captain."

Vernace studied Devlin's face. "Excuse me? Here? You're staying here? What in God's name for?"

Devlin told Vernace about the last probe. "It's critical that the launch goes as planned."

"And you're willing to give your life up for that?"

Devlin bit back his lower lip. "I'll be dead inside a year anyway. Just before the disaster struck, I was diagnosed with colon cancer. The doctor said there was no rush; if they got it out within a few months, I'd be fine. But then, well…"

"And the doc up here couldn't take care of it? I mean, I know we used to send people home for anything major, but Jansen's pretty damn good in an emergency."

"I never told him about it. Or anyone else for that matter. Even Debra doesn't know, and I'd like to keep it that way. I didn't want anyone's pity, and didn't want to put Jansen in the position of doing a procedure he had little chance of succeeding at. I've lived with blood on my hands and I wouldn't wish that curse on anyone."

"You've been sick all this time? No wonder you've been such a grump. And here I thought it was just the guilt."

"Actually, I felt fine until the last few weeks. But I can feel the pain now, feel the lump in my gut. I won't have much longer, and I'd like to die knowing I did something useful."

Vernace's forehead creased. "And you can handle this yourself?"

"I've been watching Hastings launch all the other probes. Doesn't look too hard, once it's all programmed to go."

"Good, because there won't be any help. Radio's down. Damn dust plays havoc on the antenna array and no one's been out to service it for the past few weeks—didn't see much need in wasting the manpower on it under the circumstances."

"I'll manage."

Vernace took a deep breath and exhaled. "I'll go with the second shuttle. I'll secure everyone in and tell them you're in the extra seat bolted down in the cargo bay. No one else will be back there, so no one will know until we're underway." He scanned Devlin's face, then shook his head. "I never did know how to figure you, Doc. I always admired your genius, but I didn't know whether you were an asshole or just some schmuck with incredibly bad luck." He laughed and held out a hand. "I hope that luck of yours is better when it comes to getting that probe off the ground."

Devlin shook his hand. "Thanks."

He left Debra shortly after breakfast, wanting to tell her, wanting to say good-bye, but not willing to risk having her try to stay behind with him or trying to coerce him to go. He hid out in his quarters until the second shuttle was rolling out to the launching strip, then went up to the observation deck to watch Debra and the others take off.

For the first time in his life, the solitude Devlin had always sought now felt disquieting rather than calming. Stolen moments alone in the lab or at home had always been a phone ring or door knock away from disruption. But the finality of his self-imposed isolation now sent unexpected pangs through his gut. There would be no interruption this time; no voice, unwelcome or otherwise, to interrupt his thoughts, his nightmares.

As he watched the faint dot of light reflecting off the hull of the shuttle shrink to invisibility, his first urge was to suit up and try to repair the antenna array, establish communications in case something went wrong, but he had no idea how to even start. Maybe it was for the best; maybe it would be easier to not have to explain to Debra why he never said good-bye.

The next several weeks passed slowly. He didn't mind the solitude; in fact, it was something he'd always craved. But it did eat at him wondering whether Debra would find out the reason he'd given Vernace for staying behind, wondering if she'd ever talk Jansen into letting her see the medical files and learn that he was as healthy now as the day she'd met him. And wondering if she'd understand.

When the day came to launch the final probe, Devlin hoped the automatic sequence Hastings had programmed in would send it on its way. He wasn't surprised that it did not. He was used to things going terribly wrong in his life, and that was exactly why he was still here. There was a 16 hour window to send the probe on its way before the course would have to be reprogrammed, something he had no idea how to do.

He tried to trigger the launch from the engineering room, as he'd seen Hastings do for the other launches, but it was no use. The trigger must have malfunctioned due to dust or radio failure. He suited up and went out to inspect the mechanism. He blew it clean of debris, checked the radio to make sure it still had power, then went in and tried again. No luck.

Whatever had gone wrong was something that he knew he couldn't fix in the next 14 hours. But Devlin had contemplated this moment over and over since the shuttle had left. And if his calculations were correct, the remaining jet packs, wearable units designed to help the astronauts maneuver in space, could generate enough thrust to get the probe up past the moon's gravitational pull. Once it was, the internal drive Hastings had programmed would be activated, sending it on its way.

He'd had plenty of time to modify the jet packs into a frame that could hold the probe, but there was no way to free it from the apparatus once it was up; no way other than physically releasing it by hand.

Devlin went to the shuttle bay. The enormous doors still hung open where the ships had rolled out, and the ribbed metal floor was covered with a layer of powdery gray moon dust. He donned a suit, mounted one jet pack on his back and hoisted the second one over a shoulder. As he waited for the airlock to equalize, he laughed. *Who the hell am I saving the air for?*

He made his way on foot across the expanse to where the probe was positioned for launch, then stood facing it and hoisted the spare pack up over the nose of the small probe that was only about chest-high. Securing the handles of the two jet-packs together, he created a frame around himself and the small rocket and used the modified harness system he'd constructed to fasten the rocket into place.

He smiled at the lunacy of what he was about to do, then triggered the jet packs simultaneously, lifting himself into the sky with the probe strapped to his body. The small jets struggled to lift the weight against even this meek gravity, but as they rose, he could feel the resistance abate as they picked up speed. When the fuel gave out and the burn stopped, first in the pack on his back, then a few seconds later in the other, he felt the thrill of weightlessness confirming that his lunacy had worked.

The harnesses had tightened in the ascent, and Devlin felt a brief pang of panic at the thought that, after all this, he may not be able to free himself from the probe. But he did. And the rest was easy. As he unfastened the two jet packs from each other, the probe began to drift off, and when it was clear, he flung the second jet pack in the opposite direction. There was no point in bothering with the one on his back; in zero gravity, he hardly knew it was there.

Within a matter of minutes, a small puff emanated from one side of the probe and then the other. It steadied itself in space, then a split-second burst of light from the tail end and it was on its way.

For the first time in as long as he could remember, there was nothing left for him to do. The fate of the world no longer rested on his shoulders, no longer depended on any decision he would ever make again.

As he drifted, the bright blue crescent of home shone brightly in the blackness of space and he couldn't help but wonder if Debra and the others had made it; if they would manage to survive and resurrect the world he'd forever changed. Or if he was now the last one left alive.

He turned his head and scanned the starscape, hoping to catch another glimpse of the probe that carried his last hope for redemption. Now, nothing more than a speck of light shrinking into the distance, he followed its course until his vision began to blur. And as the last of the oxygen thinned from the air in his helmet, he wondered what humanity would be like the next time around.

9

And a Time to Every Purpose

Technology and knowledge have been slowly changing our perception of disability, but individuals living with disabilities still often struggle for equality. Here's one future where the tides have turned.

It had been 6 years since Samuel Hawkins had last seen his former commander, 6 years since he'd been discharged from Walter Reed, 6 years since he'd last been able to feel anything below his neck.

The chairs lining the walls only served as obstacles to the random path he took around the waiting room, each footstep echoing along the parquet floor as he perused the artwork, mostly pictures from the old Apollo missions that he'd seen a thousand times before.

The deafening silence was finally broken by an electronic chirp coming from the oak desk sitting just outside the door to the general's office. Hawk swung around toward the young redhead who had been sitting there, silently working away on her monitor.

"The general will see you now."

He responded with a single nod.

She stood and opened the door, motioning him in.

Hawk took three steps into the modestly sized office and stood at attention.

General Richard Tarrington tapped at the screen set into the desk top in front of him. "Damn paperwork almost makes me wish for a war." Not that much real paper actually crossed his desk anymore. He gave the display one final emphatic tap, then looked up.

Hawk felt an awkward pause as the general studied him. But he didn't have to put up with this anymore, waiting for a superior officer to address him first. He'd ceased being military the day he was discharged from the spinal cord injury rehab unit. "You wanted to see me, sir."

"I thought they said it was permanent, son. 'Not a rat's chance in an ice-cold ocean,' I believe were their exact words."

"Something like that, sir. Yes."

"Then how in the hell are you standing in front of me?"

B. Aiken, *Small Doses of the Future*, Science and Fiction,
DOI 10.1007/978-3-319-04253-4_9, © Springer International Publishing Switzerland 2014

"New suit, sir."

Tarrington's left eyebrow crept up. "Doesn't look like much of a suit to me, and that sure as hell doesn't answer my question."

"Yes, sir. It does. Nanofiber suit with a Brain-Computer Interface. I'm wearing a body suit under this." He motioned to his casual beige jacket and jeans. "The material's impregnated with millions of nanotubes that slide over each other—lengthen and shorten like real muscle fibers. Whole thing's wired into a receiver in the back of the collar." Hawk turned and gave it a tap. "It picks up signals from the transmitters they stuck in my head. All I have to do is think about moving, and the suit does the rest. Even lets me feel whatever touches it, just like real skin."

"I'll be damned. When they said you had a computer interface, I thought they meant you could move a powered wheelchair around, play computer games, that kind of stuff."

"That was the first step, sir."

"Shit," Tarrington studied the man in front of him. "Why the hell didn't I know about this?"

"It's not military, sir. Private sector. And I'm their first guinea pig."

"All the better. Hell, that's what I dragged you in here for anyway, that damn brain implant of yours." He motioned Hawk toward one of the chairs across the desk from where he sat.

Hawk obliged. A screen popped up in front of him from a small emitter in the desktop.

Tarrington cleared his throat. "We first got this message from the Tranquility Base moon settlement thirteen days ago." He tapped on the screen at his desk, activating a duplicate image on the monitor facing Hawk, paused a few seconds, then tapped again. The screen remained blank, other than for the standard Marine logo in the center. "Damn thing's locked up again."

He walked over to the door. "Susan, get the IT guys up here."

"Yes, sir."

Tarrington closed the door and retreated behind his desk. "Third time this month... *computers.*" He rubbed his hands together and leaned back into the burgundy leather cushion of his high back chair. "Anyhow, the gist of it was this: Tranquility Base picked up a bogey thirteen days ago. Thing was moving so damn fast it was on them within minutes of the first blip."

"The Chinese?"

"Nah. We know what's going on over at Marius Hill almost better than they do. They don't have anything that can move like that."

Hawk's eyebrows peaked. "First contact?"

"Don't be so damn happy about it until you hear the rest. Our boys up moonside dispatched two S-17 raiders to cover T Base as soon as they picked up the bogey, tried to hail them. Damn thing shot them down before they even cleared the base perimeter."

"And Marius Base?"

"Sent up two of their own. Same results."

"Damn. And no communication?"

"Nothing. The son of a bitch just hovered there a few minutes, then headed straight for Earth. Took them all of fourteen hours to reach us. They started over California then made their way east. We had a squadron of Z-27s up in the airspace over Arizona to greet them."

"How did you keep it out of the press? Haven't heard a thing about it on the Web."

Tarrington shook his head. "That's because nothing happened. Damn thing just ignored our guys. We didn't directly engage and they didn't attack. Guess they thought we weren't worth the effort. Once it hit our atmosphere, it didn't move any faster than the stuff we've got. I don't know if they did that just to avoid attracting attention, or because they can't do any better in atmosphere."

"Maybe that's why it didn't attack."

Maybe. More likely it just didn't consider us a threat after what they saw on the moon. Knew we weren't worth the effort. We tried scanning, but it might as well have been a black hole. Everything we sent at it vanished—no return signal. We got nothing. And even though it slows down in the atmosphere, the damn thing moves like a dancer. Stops on a dime and darts any-which-way, up, down, or sideways.

"We followed it across the US, then Europe. The Chinese picked it up in Asia."

"So where is it now?"

"It stayed here eleven hours, fourteen minutes, then straight back past the moon same way it came in."

"And no attempt to contact us here, either?"

"I don't think they were here to talk."

Hawk rested his elbows on the desk and peaked his fingers. "What am I doing here, General? Why call a civilian in on this? I'd think you'd want to keep it in house."

"Wasn't my idea, Hawk." Tarrington's monitor was live again. He gave it a tap. "Send him in, Susan."

The door swung open and a thirtyish-appearing airman with olive skin and a freshly pressed uniform entered, pausing just inside the door to salute the general.

"At ease, Captain." Tarrington motioned him over to the chair next to Hawk, then called through the closing door, "Where the hell are those IT guys?"

The door pushed back into the room and his freckle-faced administrative assistant poked her head in. "May I, sir?"

Tarrington motioned at his dead monitor. "By all means."

She walked over to his desk, tapped in a series of commands and the screen sprang to life. "Just needed a reboot, sir."

"Of course. Just a reboot." He watched her walk out and waited for the door to shut behind her.

"Treats me like some kind of idiot whenever there's a computer involved." He squinched his left cheek and shrugged. "Not so far from the truth, really."

"Hawk, this is Captain David Cortez. Our first BCI pilot."

Hawk cocked his head. "Excuse me?"

"Tell him, Cortez."

"Yes, *sir.*"

"And Cortez…"

"Yes, sir?"

"At ease means *at ease.*"

Cortez shifted in his seat, appeared more uncomfortable than before. "Yes, sir." He turned his attention to Hawk. "The BCI project started up about a year ago. They equipped one of the new Z-27s with a BCI receptor designed to hot-sync with a pilot who's been fitted with brain implants. Figured that would take out all the lag time of manually controlling the aircraft. I'm the first one to receive the implants; first to test it out."

"Implants," Hawk said. "You mean like mine?" The memories, never too far from the surface, came flooding back. The dogfight over M'bangwa, spiraling toward the sprawling village, guiding his plane away from the huts…so many lives, ejecting too late for a safe landing, the sickening searing crunch in the back of his neck as he bounced off the side of the mountain. The eternity until rescue.

Cortez knew all about Hawk's accident, his comeback from paralysis to functional quadriplegic through the use of a BCI. "Exactly like yours."

"You a quad too?"

"No, I'm not."

"Then why the hell would you let someone mess around with your brain?"

"Because I got to be the *first.* Isn't that what we're all in this for? To know how Yeager felt when he broke the sound barrier. To be the first to test-pilot technology that's destined to change history. This was my shot, Hawk."

He knew exactly what Cortez meant. It was what he missed most about the Corps. There were other things in life he missed more, but the emptiness in knowing he'd never be at the controls again, never know the exhilaration of piloting a craft without knowing for sure how it would react, what it could achieve until someone like him pushed it to its limits. It was an emptiness that only another pilot could understand.

He turned to the general. "Why the hell didn't you come to me first? I already had the interface? No need to muck around in anybody else's skull."

"No fail-safe, Hawk. What if it didn't work? A pilot with no means of recovery if the computer interface goes down?"

"Too much expensive hardware to risk?"

Tarrington glared. "You know me better than that."

Hawk did. "Sorry, sir."

The general looked at Cortez. "Go on."

"It took a couple of months after the surgery before I could use the interface. A few more before I was any good at running sims. But by the time I made my first run it felt like I was part of the plane. The Z-27 maneuvers like nothing you've ever seen, Hawk." His dark brown eyes widened. "And all I have to do is think, react. The response is instantaneous."

"When the guys who tailed the alien ship reported back, it was clear from the footage that we would never be able to match its maneuverability. Sure, the Z-27 is almost quick enough, but the pilots aren't, not with manual control. And we know the Chinese don't have anything that can handle like that. The only chance we have of stopping them is with a squadron of Z-27s piloted by BCI pilots."

Hawk slowly shook his head. "OK. Again," he turned toward Tarrington, "what am I doing here?"

Tarrington fixed on Hawk's eyes. "Because we don't have a squadron of Cortezes. He's the only one with the implant."

Hawk looked over at Cortez, then back at the general. Me? You want *me* to pilot one of those things? Jesus, General. You bring me in here and hit me with *this*? You know damn well I'd be behind the controls of anything that could fly if I had a snowball's chance in hell of doing it. But look at me. You *know* I'm a frickin' quad. I can't even wipe my own ass. You said it yourself, how the hell would I recover if the computer goes down?

"There's no way in hell anyone's going to let me fly again and I've been living with that damn thorn in my psyche for five years now, sir. *Five years* trying to convince myself that it didn't really matter, that life was still worth living. Five damn years. So thank you very much for bringing it up. Thank you for dragging my disabled ass into your office to put me in my place. For pointing out just what I can't do any more, because I sure as hell can't. I can't do what you're asking, and you know it. *Sir.*"

Hawk stood and tugged his jacket smooth.

"You finished?"

"Damn right." Hawk turned toward the door and took a step.

"Stop right there," Tarrington barked.

Hawk froze. He may not have been military anymore, but old habits are hard to break. And the general's voice was, well… very general-like. He turned and faced Tarrington.

"This isn't the FAA, son. You were a damn fine pilot and there's no reason in hell you can't handle a Z-27 with a BCI. Hell, the way you move in that fancy suit of yours, you can probably fly the damn thing manually."

Hawk stared in silence. Let it sink in. Walked over to the desk and sat. "You serious? You're going to let me fly."

A raised eyebrow from the general answered his question.

Hawk was flooded with emotions, exhilaration dampened by the reality of his body. Sure, he might be able to handle a plane with a good enough BCI, but a Z-27? The damn thing was faster, more powerful than anything he had ever trained on. And what if something went wrong? How the hell would he be able to regain control? The suit let him walk, sit, eat; even play a decent game of pool. But he'd have to turn it off to hook up to the aircraft's computer, and his body would be dead weight once he made the switch.

"You can't imagine how much I want this general, but…"

"No buts, son. And it ain't just you. You're our poster boy, but we need thirty-four more just like you."

"Excuse me?"

"Intel says we've got at least a dozen of those alien buggers heading our way, and they'll be here in twenty-nine days. We can get thirty-six of the Z-27s converted for BCI control, but there's no time to put interfaces into any of our pilots. They'd barely be recovered from the surgery by the time the bogeys get here. You remember how long it took you to learn how to use the interface? We need thirty-six pilots who take their interfaces for granted, don't have to think about how the damn things work when they get into the cockpit. And right now we've got exactly two. Counting you."

"And the rest, sir?"

Our fight surgeon has pulled the database. Over seventeen-hundred quadriplegics have been equipped with BCIs in the last fifteen years. Nine-hundred and eighty-six are under the age of fifty. Sixty-six of those have some flight experience, a third of them military, and only half of those have logged any combat hours. That's eleven who have some kind of chance of pulling this off. We just have to hope the rest can survive up there long enough to at least lend a hand.

"You've got the most combat hours of any of them. It's why I called you in first; that, and the way you handled yourself in Africa. You're a natural born leader, son. And that fancy suit of yours makes me feel even better about my decision. Maybe our engineers can rig up some way for you to reconnect with your suit if something goes wrong up there. The rest of them…they're all stuck in wheelchairs or bulky exoskeletal suits that'll have to come off just to fit them in the cockpits. For them, it's heaven or hell."

"Why just the quadriplegics? Why don't you use the paras. At least they have some use of their arms. They'd have some chance of surviving if the interface goes down."

"Different BCI. They have less than half the cortical chips that you do. Only enough to control their legs. The tech guys say the array of microchips in the paraplegics' heads aren't comprehensive enough to control a Z-27's computer. It's got to be quads."

Hawk knew what life was like without the suit. He couldn't even take the damn thing off by himself. Sari had to peel it off him every night to wash him. He wouldn't even sleep without it. When it was off, he was helpless, vulnerable in a way that no man could imagine unless he'd lived through the same hell. At least he'd had Sari, God bless her. Most weren't lucky enough to have found love and kept it, something that's never easy, something that, like everything else in life, is a hell of a lot harder once your brain becomes detached from your body.

"And you want me to talk them into this insanity?"

Tarrington shook his head. "Already done. And it wasn't too damn hard. These boys and girls are dying to get up there again…maybe not the best choice of words, but they *are* willing to die up there, just for the chance to fly again. I thought you'd be too."

"Oh, I am, sir. More than willing. I just don't want to be responsible for dragging anyone else along. And," he looked the general in the eyes, "I'm scared shitless."

"You? Hardly fits your profile."

"Not about the aircraft, sir. And sure as hell not about dying. I came to terms with that years ago. I'd rather die in combat than any of a hundred other things that could happen to me, especially since, well, you know." After a brief silence, a wry smile crept across Hawk's lips. "What scares me is having to tell my wife."

*

Sari Gandarhi was the nurse assigned to Hawk the day he was wheeled into the Walter Reed Rehabilitation Center.

After two and a half weeks of being poked and prodded on the neurosurgical ward, he was relieved to finally be moving on to the next phase of his recovery. It only took one physical therapy session to learn that the easiest part of his recovery was behind him.

Despite the best efforts of one of the top spinal cord injury teams in the world, Hawk remained nearly helpless. His limbs were like jelly except for the moments when they went into uncontrollable spasms, eliminating any precarious balance he may have been able to achieve once propped up in his chair. Eventually, they got the spasms under control and taught him how to move a wheelchair around by barking voice commands into a computer that controlled the drive mechanism.

Most men would have retreated into what was left of their bodies, shut out the world. But Hawk was determined to get his life back. He fought off the overwhelming loss of self-esteem his useless limbs forced upon him with a sense of humor that contradicted his situation, a coping mechanism that resulted in a parade of mental health professionals—social workers, psychologists, psychiatrists, all determined to help him deal with reality. It was one of those psychiatrists who made the recommendation to BioSynthesis that Sam Hawkins would be the perfect candidate to test their new nanofiber suit. It wasn't hard to convince Hawk to agree, and the cerebral cortex implants were placed the following week.

When he was finally discharged home, Sari took the private duty job of helping him reintegrate. He wanted so much more, but there was no way he'd risk scaring her away by telling her how he really felt.

The technology incorporated into the new suit required a cerebral interface far more complex than any previously implanted, but with Sari's help Hawk was able to master it within a matter of weeks. His self-esteem grew with each day that he became more independent, and he started to relate to Sari on a more personal level. It wasn't long before he realized she felt the same way about him as he did toward her. When he finally mustered the courage to venture out of the house alone, his first stop was to pick up the diamond ring that would signify the next phase of their lives.

He worked tirelessly for BioSynthesis over the next several months, work that gave his life meaning. But it wasn't flying. There was nothing like piloting a fighter plane into action.

He told Sari of Tarrington's offer as they cleared the table after dinner.

"What do you mean they want you to fly a mission? They don't own you anymore."

"I *need* this, Sari. I thought I was past it, but I realized today that I'm not."

"Are you crazy? I am *not* going to lose you. Not now, not after everything we've been through, everything *you've* been through. I'm not going to let those selfish bastards have you again."

Hawk took Sari's hands and guided her to the sofa, sat down beside her and lifted her chin until her bright hazel eyes fixed on him. "It's what *I* want, Sari. Not what they want. It's what I'm trained for, what makes me feel alive."

As she studied his eyes, the anger slowly drained from her face. She bit back her lip and nodded gently. "OK."

*

Hawk took off for the base in New Mexico the next day, where a dozen Z-27 sim units had been set up in a facility near the hangar and barracks where the BCI pilots would be training.

The process for transferring BCI control to a sim unit was the same as it would be when the time came to take the controls of a Z-27. They strapped him in, disconnected his suit and swung the helmet over his head. The controls sprang to life instantaneously, but before he even began the preflight check list, the flashbacks started. The African Wars, the spiraling plane, the crash. With the suit deactivated, he couldn't tell where his hands or legs were, couldn't move or feel a thing. He hadn't felt like this since he'd first put on the suit, except for the brief moments in bed with Sari at his side. His heart raced and the breaths were coming faster, shallower.

The sim tech stopped the program 90 seconds in and called for the flight surgeon.

But Hawk regrouped quickly and convinced the tech to cancel the call and restart the sim.

This time he was ready. He connected with the plane, and the sky became his playground. Before he could even finish thinking *bank left,* it happened. So natural, so untethered. It was what Hawk was born to do.

The first sim was scheduled to be a simple take-off and landing, but the tech couldn't get Hawk off the program. He banked toward the CGI-generated Gila National Forest hanging low over the tree line, then over toward Albuquerque and up to Santa Fe. It felt so real, so right. He set course for Taos, hoping the program would give him a glimpse of the crowded ski slopes, but the fuel warning light flashed on his visor, and he reluctantly headed back to base.

Once he got comfortable with the interface, the tech taught him how to use a fail-safe they'd rigged up for him. In case of BCI failure, he could re-engage his nanofiber suit with a pattern of left shoulder shrugs, the only active motion below the neck he'd managed to regain after months of rehab. Two shrugs, pause, one shrug, pause, two shrugs. It triggered a switch that reactivated his suit. The same pattern with the right shoulder triggered the ejector seat.

He had nearly a week to train before the rest of the recruits settled in at the training facility, and spent every moment he could in the sim unit.

*

The chatter in the hangar where the flight recruits had been assembled waned as Hawk walked in and made his way to the front of the group. About two thirds were in wheelchairs, the rest in exoskeletal devices, clumsy precursors to Hawk's suit. But they'd all made it there by themselves, a testament to their proficiency with the BCI units that let them interact with the world.

He turned to Cortez. "Everyone here?"

"All accounted for."

Hawk scanned the room. "Thank you all for coming. Sorry we couldn't tell you more before you hauled your asses all the way out here, but we knew you wouldn't pass up the opportunity to fly again."

"And just what would *you* know about *us*, Captain?" The voice was bitter and infused with the raspy quality of tobacco-abused vocal cords. It came from one of the wheelchair-bound men in front. "How much time have *you* spent chained to a chair?"

The room rumbled in concordance, but one man standing in the back spoke up before Hawk could respond. "Who cares, man. They're going to let us frickin *fly* again."

There were scattered shouts. Some in agreement, some not.

Hawk held up a hand to quiet the room. "No, it's a fair question." He looked at the man in the back who had spoken up in support. "And don't be so quick to jump on board, because if you agree to this there's a damn good chance it'll be the *last* thing you'll ever do."

The room buzzed again, and Hawk spoke louder. "But I'm willing to bet not one of you is going to turn me down."

"Bullshit, man." One of the chair-bound men spun himself toward the door.

Hawk's voice sharpened. "You really going to pass up a chance to take a Z-27 into combat?"

The wheelchair stopped, turned back toward Hawk. "The Chinese?"

A grin crept across Hawk's face. "Nope." He proceeded to brief everyone on the mission.

"So *we're* the sacrificial lambs…*again*."

"Looks that way."

"You're sure mighty anxious to send us all to hell. You think our lives are that pathetic just because we're stuck in these chairs? You don't know crap about us, man."

"I *am* one of you."

The man studied Hawk intently. "You the suit?" Rumors about a quadriplegic successfully using the first nanofiber suit had been bouncing around the web for weeks, but most of the blogs said it was bogus. Just more false hope.

Hawk opened his shirt and pulled on the blue stretch nanofiber.

"You look like friggin Spiderman." He laughed and wheeled around seeking the approval of his cohorts.

Hawk laughed too. "You got a name?"

"DefCon."

"Excuse me?"

"Juan Defortenzio-Contreras, in those files of yours. But there are a million Juans out there and you gringos always butcher my last name. My CO shortened it to DefCon. Kind of stuck."

"Works for me." Hawk remembered the files he'd studied. "Navy, right?"

The man nodded. "USS Reagan. Eighty-seven missions without a scratch."

"Then how'd you end up in that chair?" It was only then that Hawk noticed the man was missing both legs.

"Got flipped in a rough weather landing. By the time they scraped me off the deck, I'd lost half my blood. They took both legs in sick bay. It wasn't until I came-to a couple days later with two dead arms that they figured out I'd snapped my neck too."

The room fell silent. Gruesome story, even for this crowd.

"Well, we could sure as hell use you. You've logged more combat time than anyone in this room, besides me. But if I can't count on you to watch our backs up there, I need to know now. You're no good to us if you haven't gotten over it yet."

Every man and woman in that room had had to deal with *it* after their injuries; had to conquer *it* before they could to move on with their lives.

DefCon fixed on Hawk's eyes. "Don't try and get in my head, man." Then a smile slowly broke across his face. "So when do *we* get those suits?"

"This one's just a prototype, but I've already talked to the general. He'll do his best to see that anybody who makes it out of this alive gets bumped to the top of the list, once the suits hit the market. And Uncle Sam will foot the bill."

The room filled with hoots and whistles. DefCon spun around and nodded to accept the thanks.

By the time the din settled, the mood in the room had eased considerably. Hawk worked his way through the group, putting a face to the name of each man and woman he'd signed off on to join the team. Every one of them had a story, but they weren't here to tell it. They were chosen to be here because they'd moved past their stories, and Hawk knew just how hard that was.

Hawk's difficulty with the first flight sim proved to be the norm rather than the exception. Of the 46 pilots recruited, only six couldn't get past the panic attacks precipitated by the change-over of the BCI interface from chair to plane.

For those who were able to make the adjustment, the schedule was rigorous. Over the next 3 weeks, they each logged over 100 hours of flight sim time. By the time the Z-27s were all fitted with BCIs, they were ready to fly maneuvers. They divided into groups of seven, one alternate to each squadron, with Cortez or Hawk taking each group up until the pecking order sorted itself out. DefCon proved his worth and by the end he and three others joined Hawk and Cortez as the six group leaders.

They were off the sims and running war games against each other for about a week before word arrived from Tranquility Base that the bogies were on the way.

"Well," General Tarrington shook his head as he started to speak, "I never thought I'd be saying this, but you bastards can fly circles around any other squadron I've got." He met the eyes of each flyer as he panned the room. "And it's a damn good thing, because Tranquility Base has confirmed what we've been tracking on our long-range scans."

"They didn't bother with the moon bases, this time around. You all have just under seven hours before you're deployed, so get back to your barracks. Go say what you've got to say to your loved ones, but no alien bullshit. We're keeping it on the QT as long as we can. And get some shut eye. You're going to need it. Reveille's at oh-four-hundred. Meet back here to pick up your assignments."

Tarrington saluted and walked out as they all sat motionless.

"Right," Hawk said. "You heard the general. I know none of you are going to get much sleep, but get some down time. Hook in your Sound Buds, put on your VR goggles and hit the beach, whatever it takes. But get your mind off all of this and try to get some rest."

Hawk and Cortez stayed until the last of them had left.

"Think they're ready?" Cortez asked.

"For a dog fight with aliens that downed two of our moon flyers in less than eight seconds? Who the hell could be ready for that?"

"Thanks, Hawk. I'll rest a lot easier now."

"But they're not going to crack up there. They've already faced their mortality. Each and every one of them. They're ready, Jav."

Hawk gave Cortez a pat on the back. "Let's get some rest."

*

Sari was still up when the phone rang. She'd spoken with Hawk at dinner time, just like she had every day since he'd been away. When she saw his number come up on the screen again, she couldn't help but think the worst.

She tapped the phone on. "What's wrong? What happened?" It was a relief to see her husband's face, serene as always, on the small bedside monitor.

"Everything's fine, hon. The wait is finally over. We're on for tomorrow."

"And that's supposed to make me feel better?"

"It means it'll all be over soon. I should be back home in a few days."

"I was hoping they'd just call the whole thing off."

Hawk laughed. "Wasn't our call."

"Then go kill the bastards who took you away from me and get your ass back here."

"Aye, Ma'am." He saluted.

"I love you, Sam."

He smiled. "Nothing's going to keep me from getting back home to you."

She forced a smile through her tears as she disconnected.

*

Reveille rang out at four sharp and they were all heading for the hangar within 15 minutes. Not one of them had come off their BCIs for the night. They reclined in their chairs, lay down with their exoskeletons on, tried to rest. But the synthesized bugle sound was like a release from prison more than a call to duty.

Tarrington was waiting when they filed in.

"Looks like they're splitting up into two groups. Too soon to know where they're headed yet, but it's a good bet they'll target the most industrialized areas first."

"Asia and us," Cortez said.

"Right. The Chinese are covering Asia. Hopefully they can hold their own, at least take a few out before the buggers head our way."

DefCon's voice rang out from the group. "They got BCIs too?"

"Doubt it," Tarrington answered. "Probably don't have any virtual decoys, either." They'd all been briefed on the new technology, a missile-lock jammer that shifts the energy signal and mass detector 15 degrees off starboard, then varies the shift every 10 seconds. "Not sure if it'll work against the buggers, but it may give us an edge."

"Christ," DefCon muttered. "The Chinese'll be sitting ducks."

Cortez grunted. "First time I wished they had something we don't know about."

"Not likely," Tarrington said. "Cortez, you take platoons one, three and five. Head over to Langley. We want Washington covered. Get the fighters down for refueling and get your people out of those cockpits for some down time before they have to engage. You should have a couple of hours there before we know what their final deployment pattern will be."

"Hawk, you've got the other three units. We're keeping you right here for now. You lucky bastards get a few extra hours of R&R, so get back to the barracks."

"If it's all the same to you sir, we'd rather stay here. We're getting cabin fever over there."

Tarrington scanned the eager faces.

"All right. Just stay the hell out of the planes until you're ready to deploy. I want everybody fresh when the time comes."

"Aye, sir."

*

The hours passed slowly. Most of the pilots huddled around a large monitor at the makeshift command post following the progress of the attack force, breaking the monotony with occasional trips to the coffee machine. Several attempts at communicating with the aliens proved to be as futile as they had during the first encounter. At 10:45, all 12 ships slowed just outside Earth's atmosphere. At 10:48, six of them darted down over Hong Kong.

Within minutes, a squadron of Xiangs headed to intercept. The ensuing battle was over in minutes, with 67 Xiangs downed. They'd only managed to disable one of the alien ships, which exploded into dust before it hit the ground, the apparent victim of a very effective self-destruct mechanism.

The whole thing played out in high-resolution satellite images as the BCI quad pilots looked on. The attackers strafed the city setting off a series of aerial blasts, and Hong Kong went dark within seconds. Not even a sign of a back-up generator still functioning.

Cortez's head swung slowly from side to side. "Jesus. What the hell was that?"

Tarrington's phone chirped and he tapped it on. "Uh huh…right…got it." He disconnected.

"Just what it looked like—some sort of E-M pulse bombs. Shut down the whole damn power grid. They're making their way across Asia now."

Hawk motioned toward the monitor where a panned out image in the top right corner of the screen showed the position of all the alien ships. "And the rest are heading our way."

*

The first wave to hit the US came over Washington, D.C. The aliens had split their remaining ships into two groups of three, no doubt buoyed by the easy slaughter over Asia.

Hawk was already airborne, watching the progress on his monitor while leading his own squadron to intercept the group closing in over California, about 15 minutes behind the first wave.

Cortez's squadron managed to evade the first salvo of missiles and take out one of the alien ships before the initial pass was over.

"*Nice, Jav.*" Hawk's nerves began to settle. Those bastards *were* vulnerable. He accelerated toward his target. "Let's go get 'em, boys."

"Hold your position, Hawk!" It was Tarrington. "You seeing this?"

Hawk glanced back at his monitor. The second pass over D.C. had taken out every ship except Cortez's. Four of the pilots had ejected, the rest were gone.

"Jesus! What the hell happened, Jav?"

"They took out our BCIs. Must not have been whatever it was they blacked-out China with; anything that big would've taken out all our avionics. Probably theirs too, if they used it in close quarters. Must have some way of generating a low-level pulse they can target in a dogfight, just enough juice to fry the micro-current circuits in the BCI. Everything else seems to be intact. Had to switch to manual." An option none of the other pilots had.

Cortez was tailing the alien ships, barely keeping pace.

"Get the hell out of there," Tarrington barked. "You're no good to us dead."

Cortez held course for a few more seconds, then turned back to Langley.

"Our turn," Hawk yelled. Despite slowing his squadron at Tarrington's order, the western wave had closed on their position and was now in weapons range. "Engage decoys. We don't know what they going to throw at us first. If your BCIs go down, don't wait around to see if they'll reboot, just get the hell out of the plane."

He split the group off in pairs. No point in lining up like ducks. Hawk and DefCon each took a second flyer and closed the distance, approaching the group of oncoming ships from either side. "Lock and fire!"

They clipped two of the ships, but barely slowed them down. There was no return fire.

"Maybe we knocked out their guns," DefCon barked.

"Focus," Hawk yelled. "We probably just aren't in range for their EM pulse yet. Keep your distance, and maybe we can survive this."

But the three attacking ships had already split up, closing the distance quickly. DefCon was the first to spin out.

"Eject!" Hawk ordered.

"Hell, no." He aimed his rolling plane in the direction of the attacker, who easily evaded the attempt to ram. DefCon slammed into the foothills of the White Mountains a few seconds later, engulfed in flames.

"Pull out!" Hawk barked, but two more of his squadron had gone dark. He could only hope his pilots had gotten out. The two remaining ships followed as he retreated from the attack.

"General, you following this?"

"Keep moving, Hawk," Tarrington answered. "Backup is on the way. I'm sending up everyone we've got left. We'll swarm the bastards."

"You're sending them into a massacre, sir."

"We've got to take those buggers down, Hawk. No matter the cost."

"Then let's at least give our guys a fighting chance." Hawk turned his plane back toward the enemy and the two remaining pilots in his squadron started to follow. "Hang back and wait for the rest."

"No way, Hawk. We're coming with you."

"Appreciate the sentiment, but that's an order."

They hesitated, then dropped off.

"Speaking of orders," Tarrington barked, "what the hell are you doing, Hawk? I said back off and wait for the squadron."

"Look, on that last pass we each got off a few shots before they returned fire with that EM pulse. Same for Cortez?"

"Yeah," Tarrington said. "What of it?"

"There's no way those bastards waited for us to fire first out of some moral pang, not the way they massacred our raiders up on the moon."

"What's that got to do with anything? They were probably just ignoring us until we became a nuisance. Swatting at flies."

"Maybe, but I'm willing to bet that EM rifle of theirs is short range only. If we can figure out exactly what that range is, we can send the rest of our fighters at them from a safe distance."

"Good thought, Hawk. I'll have the tech guys analyze the data from the two dogfights, see if they can extrapolate the range."

"No time. They're heading for LA now. Too many lives, sir. I'm going in hot. I'll send back data as I go. Track me until I go dark, then mark the distance and send up everything we've got."

There was a brief pause before Tarrington answered. "Crazy bastard. You sure you can activate that suit of yours and go manual if you get hit?"

"No sweat."

Hawk knew his suit's interface would be every bit as sensitive to an EM pulse as the Z-27's BCI, probably more so. The whole damn thing ran off micro-currents and didn't have any of the shielding that safeguarded the main systems of the Z-27. But what other choice was there? The piezoelectric switch that triggered the ejection seat should still work. Hell, some of the other pilots had been able to eject, so it *must* work. Or were those the ones with partial upper extremity function that could have hit the manual eject button?

Didn't matter now.

Hawk was closing within a half mile of the bogies. He had five missiles left. He fired the first.

The buggers ignored him.

He fired the second, the third. Five-hundred thirty-seven meters and closing. Just after the forth shot, he felt it. Oblivion. Disconnection. No sense of his arms, his legs. Sensory underload. Nothing but his vision, blurs of light as he spun toward Earth.

Images of Africa raged through his mind.

Disoriented. Alone.

Panic descending on his untethered psyche. His heart pounded and he gasped for air, reached for his helmet, but he had no hands.

Two Gs pressed his limp body back against the seat as he fought to try the improbable. Who'd have guessed a simple shoulder shrug could feel like bench-pressing an elephant.

Two…One…Two…

No response. His suit was as dead as the plane's computer.

Then a peaceful calming voice suffused his thoughts, subduing the panic. *Let it go, Hawk.* It felt so right. *Let it go.*

He relaxed back, succumbing to the one answer he'd sought a thousand times before. Ready to finally give up the fight that had kept him alive, brought him back from the oblivion of Africa.

"Get out of there, Hawk." Tarrington's voice seemed distant, unimportant.

Hawk smiled. He was young again. Heading for home base, caught in a run down with no way to make it safely past the catcher, then seeing the opening that would get him back to third. *Come home.* It was Sari's voice. He looked back over his shoulder toward the umpire and saw the face that made his life matter. *You promised, Sam.*

"Get out of there, Hawk." Tarrington, again. "You've got fifteen seconds. Eject."

Sari.

The sweat poured down over his eyes as he fought against the harness restraining his right shoulder.

Two…One…

The harness locked. *Shit!*

"Now, Hawk!"

The Z-27 hit an air pocket, jerking Hawk hard to the right, creating some play under the harness.

"Five seconds. Get the hell out of there, God damnit!"

Hawk clenched his teeth, focused.

Two…One…Two…

*

"He's moving." It was Sari's voice. "Hawk?"

He opened his eyes. Sari was on his left, Tarrington and Cortez on the right.

"Glad to see she hasn't killed the both of you." Hawk smiled.

"I thought about it," Sari said. "When they told me what you did, I was ready to kill you, too. What were you thinking?"

Hawk looked at Cortez, who gave him a warm smile. "Glad to see you made it."

"Told you I could fly circles around you."

"Try doing it with the BCI disengaged and both hands tied behind your back."

"Touché."

Hawk turned to Tarrington and let his eyes ask the question.

"Got 'em," he said. "Thanks to you, we got every last one of those bastards."

"Me? What the hell did I do besides get my ass shot down?"

"We used the data you sent back; set our attack distance to stay just out of range of their EM weapons. The rest of your pilots took them out without another single casualty. If you hadn't figured it out, we'd still be in the dark."

"Literally," Cortez said.

Hawk eased back into his hospital bed. "The doc say when I can fly again?"

Sari cleared her throat. Loudly. "Would you gentlemen excuse us?"

Cortez had an ear to ear grin. "Oh, you are in *so* much trouble."

Part II

The Science Behind the Fiction

10

The Invasion of Modern Medicine by Science Fiction

The allure of science fiction, like beauty, is in the eyes of the beholder. Yet, a glimpse beneath the surface of good science fiction is often a glimpse into the future of scientific discovery.

Like many fields of science, the future of medicine is frequently predicted by the science fiction writers of today. And while it's fascinating to speculate about what the future of medical science will bring, it is perhaps even more interesting to look back and see how today's medical marvels may have been influenced by yesterday's dreamers.

Many of today's medical advances were presaged by science fiction stories of the past. Is this because those writers had an uncanny ability to predict the future, or was it because they had a gift for stimulating the imagination of children who would dare to become the future generations of scientists to make those dreams come true? Most likely, reality lies somewhere in between, but either way it's hard to argue the importance of the vision of the most gifted of science fiction writers; authors like Jules Verne, Arthur C. Clark, Isaac Asimov, and Gene Roddenberry to name too few, have influenced generations of brilliant scientists who have shaped our present and are in the process of shaping our future.

It is at best difficult to trace the roots of invention, the original idea that eventually leads to a new discovery. Inventions rarely occur in one giant leap, but rather tend to occur in a series of steps, each painstaking advance built upon the work of those who came before, until the final step—the breakthrough that leads to a fundamental change in our understanding of how things work. The scientist credited with the ultimate breakthrough is often unaware of the motivations of the scientists who came before, those who laid the foundation for the work. Therefore, in many cases we can only guess at the motivation behind the world's greatest discoveries.

With that in mind, let's take a look at the science behind the stories from a historical perspective about where some of these ideas may have come from, as well as doing a little speculation about where continued scientific development in these areas may lead us in the future.

B. Aiken, *Small Doses of the Future,* Science and Fiction,
DOI 10.1007/978-3-319-04253-4_10, © Springer International Publishing Switzerland 2014

Robotics—Artificial Intelligence

Stories:

If He Only Had a Brain

Perhaps creating the quintessential image of modern science fiction, Isaac Asimov gave form to imagination with his depiction of Robbie the robot in the short story *A Strange Playfellow*, in 1940. Not only did he bring to life the physical form of an artificial intelligence being, but more importantly, in his later writings he defined the basic principles on which many subsequent books about robotics would be founded: the three laws of robotics. These laws defined how robots would interact with humanity while, even with their superior strength and intelligence, they would never be a threat to humanity. In defining the basic make-up of a robot, Asimov allowed his readers to embrace the concept of artificial intelligence instead of fearing it.

Our fascination with robotics dates back to a time long before Asimov popularized the term. As far back as circa 350 BC, a Greek mathematician named Archytus of Tarentum built a steam-propelled mechanical pigeon. Although the scientific limitations of the day limited the rate of technological development, the idea of mechanical creatures persisted, and in 1495 the ever-prescient Leonardo DaVinci built an armored knight that could move as if there was a real person inside of the suit.

Mechanical automatons remained popular through the centuries, but it was not until Czech author Karel Capek's 1921 play, entitled *Rossum's Universal Robots*, that the term 'robot', derived from the Czech word 'robota', meaning compulsory labor, became a part of our culture. Five years later, the movie *Metropolis* introduced robots to movie goers around the world.

But it was Isaac Asimov who is largely credited with bringing the terms 'robot' and 'robotics' into the mainstream of the literary world, starting with his first robot story in 1940, and widely popularized in his compilation of short stories in 1950 entitled *I, Robot*. It was here that the seeds were sown that would forever alter the way we interact with our environment, for with the dawning of the computer age the synergy of automated machines with artificial intelligence would become reality. Future generations of scientists who were raised on the imaginings of Asimov were determined to make their dreams come true.

By 1956, the term 'artificial intelligence' was introduced into our culture, as the Dartmouth Summer Research Project on Artificial Intelligence accelerated the transition of robotics from the realm of science fiction to the reality of hard-core science. The advances in entertainment and industry that followed over the next half-century have been widely touted. These days, no one thinks twice about working with an interactive computer or seeing robotic arms in an automotive assembly plant, but machines with the capacity to think and interact with their environment are still quite basic in comparison with the human brain.

As we get closer to being able to mimic the human brain, closer to the ability to build a machine that looks and thinks like us, at what point will the difference between human and machine be indistinguishable to the casual observer? And what will that mean for the future of humanity?

Robotics—Telerobotic Surgery

Stories:

Done That, Never Been There

The utilization of robotics in medicine is less mature than in industry because of the constraints of efficacy and safety, not to mention complexity, in dealing with the human body, but has already started to revolutionize the way health care is provided.

Computerized machines with robotic arms and tools allow surgeons to operate through tiny openings, minimizing tissue damage; and by linking to these machines via an internet connection, surgeons can now operate on patients in distant locations. A surgeon with a particular skill may not be available in a rural town, but a robotic surgical device can be set up in that small town hospital, where it can be remotely controlled by a surgeon many miles away. This is not a mere whim of science fiction, but is already reality. On September 7, 2001 a doctor in New York performed gallbladder surgery on a patient in Strasbourg, France using a computer and an internet connection to manipulate a robot on the other side of the Atlantic [3]. And in 2003 the world's first telerobotic remote surgical service was set up in Canada, enabling surgeons at a teaching hospital to assist in surgical procedures taking place at a rural hospital more than 400 km away [3].

Though not yet in widespread use, this technique holds the promise of allowing surgery not only in rural areas, but even aboard ships—on the ocean, in outer space—and perhaps on the first Mars colony. Emergency medical care will be within reach even when the doctor is not in.

Robotics—Exoskeletal Devices

Stories:

A time to Every Purpose

Locked In

External bracing has long been used to help support paretic limbs, helping people to become more functional. For the most part these have been static braces. Forty years ago, metal stirrups bound to the paralyzed limb with leather bands was the norm; sometimes movable joints would be incorporated at the knee or ankle to allow more natural motion. These devices were cumbersome and provided only passive support, for the most part. Spring-assisted joints were sometimes employed, but only closed a small portion of the gap separating the function of the brace from that of a natural joint controlled by healthy muscles.

The development of new flexible plastics, and later light-weight materials such as carbon fiber, helped orthotists (brace makers) design more comfortable and more functional braces, but the designs have basically remained passive—at best energy-assistive with spring-action across joints.

The melding of biotechnology with the art of brace making over the past couple of decades is giving birth to an entirely new type of bracing, the so-called 'wearable technology' of powered exoskeletal devices. One of the first designs successfully developed to help paraplegics walk again, an Israeli device called the ReWalk [4], is an elaborate bracing system that supports the lower half of the body and incorporates motorized joints at the hips and knees. Batteries enclosed in a backpack provide power for preprogrammed computerized sequences of motion such as standing up or walking, which are modulated with motion sensors; the wearers use a wrist band to determine which sequence the computer will run, then triggers the motion to begin by shifting their body position.

Numerous similar devices are in various phases of development around the world, such as the Ekso (Bionic Exoskeleton, Berkeley, California), HAL (Hybrid Assistive Limb, Japan), and the REX (Robotic Exoskeleton, Australia). A multitude of materials and designs are being explored in an attempt to find the best way to achieve the goal of helping people with paraplegia or quadriplegia to walk again.

In my story *A Time to Every Purpose*, the use of a nanofiber suit made of microscopic fibers that slide over each other to contract and relax like muscles provides a comfortable form-fitting device that can be worn under clothing; the ultimate exoskeleton that will be so lightweight, close fitting, and effective that no one but the wearer will know it is being used. This level of sophistication is likely decades away at best, but is what researchers today are striving towards.

Bionics and Brain-Computer Interfaces

Stories:

If He Only Had a Brain

Locked In

A Time to Every Purpose

Freudian Slipstream

In 1974, the word bionic became a part of American pop culture, thanks to the introduction of a new TV hero, Steve Austin, *The Six Million Dollar Man*. The show appealed to action fans and science fiction fans alike. 'Bionic' is actually an amalgam of the words biology and electronic, and refers to using what we know about how our bodies work in order to build synthetic replacement parts. The goal is to be able to replace a defective body part by manufacturing an electromechanical equivalent for that part. In theory, we should be able to exactly reproduce the function of any body part with artificial components that won't age and wear out as quickly as the original parts we were born with. In fact, we should be able to even improve on the design, or as Steve Austin's boss said, we can build him "better than he was before. Better, stronger, faster."

Although bionics seemed like pure fantasy to most of us who were glued to our TV sets each week back in the 1970s, the field of bionics has become reality faster than most of us would have imagined. One of the first successes in this area was the cochlear implant, introduced in 1991. This tiny bionic ear can allow some people with what used to be permanent hearing loss to hear again. Essentially, it recreates the ear's ability to send electrical impulses, representing sound, to the brain.

The human ear has two main parts—the outer ear and the inner ear. It works by converting sound waves to electrical impulses that the brain can understand. The outer ear picks up sound waves and sends them to the inner ear, where a small, snail-shaped organ called the cochlea converts those waves into electrical impulses, which are sent directly to the brain.

Designed to mimic the function of the ear, a cochlear implant also consists of two main parts: an external part (like the outer ear) and an internal part (like the inner ear) which is surgically placed into the ear. The external device consists of a microphone, a sound processor that converts the sound to an electrical signal, and a transmitter that sends the signal to the implanted device. The internal device (like the inner ear) consists of a receiver that picks up the signals sent by the external transmitter and sends them to tiny electrodes implanted by a surgeon into the cochlea of the inner ear, next to the tiny nerve fibers that go back to the brain. From there, the impulses are sent to the brain along the same pathway that is used by a normal healthy ear.

Current research is focusing on a much more daunting task: an artificial eye. Our eyes work by focusing an image on the retina, a small area in the back of the eye that has thousands of rods and cones, tiny receptors that convert the image into electronic impulses which are then sent to the brain.

Several designs for an artificial retina have been developed. Though not as successful as the artificial ear, some of these devices have restored rudimentary vision (light and shadows) to people who were blinded by disease many years earlier. One of the first clinically successful designs in this area came in 2000, from two brothers from Illinois, pediatric ophthalmologist Dr. Alan Chow and his brother, Vincent, an electrical engineer who designed a chip that successfully restored limited black and white vision to six patients with Retinitis Pigmentosa, an inherited degenerative disease of the eye's retina. Two of these six patients were a pair of twin brothers who, though close, had not actually seen each other for years [7].

There are several basic designs being studied. One utilizes a camera which sends signals to a small chip implanted in the back of the eye, while another uses photoelectric cells embedded directly onto the chip itself. In both cases, the chip converts the light signals to electrical impulses, which then travel back through the optic nerve to the brain. Another approach had been to send light signals from a camera directly to a chip embedded in the occipital lobe, the part of the brain that interprets vision. The progress has been remarkable, but none of these devices can restore perfect vision—yet.

Another area of research that has yielded promising prototypes is in the fabrication of prosthetic limbs, where medical research is currently making the transition from prosthetics (passive artificial devices that replace a lost limb) to bionics (limbs that are controlled by the human brain). Traditional prosthetic legs allow people to walk and prosthetic arms and hands allow people to manipulate objects, but both are limited in their effectiveness; none of today's designs comes close to actually recreating the real thing, but new bionic limbs hold the promise of doing just that.

A prosthetic arm like the one Luke Skywalker received after his arm was cut off by Darth Vader in *The Empire Strikes Back* seemed like pure fantasy when we first watched the drama unfold on the silver screen in 1980. But that kind of technology may not be too far in our future. We already have many of the engineering skills to build such an arm, but making it part of the human body is quite another thing. Current research is working on various ways to accomplish this.

Myoelectric prostheses have been studied for years; they work by attaching electrodes to muscles in the residual limb. The wearer can then activate motors in the limb by flexing those muscles. A more recent advance, not widely available yet, is to use signals directly from nerves in the residual limb; the electrodes are connected to the nerves rather than the muscles. This creates a more direct path to the brain and more closely approaches the way we control our real arms and legs.

The newest technology in this area is focusing on highly refined brain-computer interfacing; in other words, a way to hook the artificial limb controls directly to our brain, so that we can move the mechanical arm with just a thought or even a subconscious gesture, just like we do with our natural arm. The only thing a prosthetic arm would lack at that point is sensation, something critical if the prosthesis is to allow us to truly interact with our environment in a natural way. This problem, in theory, can be solved by implanting sensory receptors into the limb that will send signals back to the brain, enabling the wearer to actually feel what the limb is touching. Once we learn how to make these connections, a prosthetic limb should finally begin to approach the OEM (original equipment manufactured) part; we can all be like Luke Skywalker—sort of.

Brain-computer interface work has now progressed to the point where quadriplegic patients can control computer cursors, robotic arms, and even a wheelchair via a computer chip embedded on the surface of their brain, which is connected to a computer that can interpret their thoughts. Though still in relatively early experimental stages, this work is progressing rapidly and holds the promise of helping people with paralysis regain control of their lives.

The Ethics of Cloning and Biological Warfare

Stories:

The Last Clone

Once, on a Blue Moon

The idea of human cloning has long been a staple of science fiction, and from the time the idea was hatched (pun intended), the ethical debate over whether we should began to outweigh the scientific debate over whether we could. One of the earliest examples, Aldous Huxley's *Brave New World* presented us with a world in which people were mass produced to fill specific roles in society, to populate a predestined caste system. The very idea of this tears at the mind of anyone born into a free society.

Advances in technology have lent stories like *Jurassic Park* an air of authenticity, and the true science behind cloning eventually led to the birth of Dolly, the first mammal to be successfully cloned by transferring the nucleus of a cell from an adult (a cell from a mammary gland, in this case) into an unfertilized oocyte (egg) [29]. She lived for 6 years before succumbing to lung disease. And with Dolly came the first true hope (or threat, depending on your point of view) that cloning a human may be within the realm of possibility in our lifetime.

Also with Dolly came the rush of politically charged ethical arguments about how and why cloning should be regulated, a debate which continues to this day.

In my story, *The Last Clone*, the issue of *can we* has been leap-frogged, the issue of *should we* is alluded to, and the issue of *Well if we can, but if it's really, really expensive, who should have the right to do it?* becomes the center of the story. And the story, which also revolves around the incredible potential of nanomedicine, changes focus to the problem of how to equitably ration health care when our technology has led to a system where it's too expensive to give everyone the very best.

In *The Last Clone* the solution is simplified and we all get to cheer a little when the uber-rich guy who thinks he's better than everyone else is dealt a dose of poetic justice. Ahh, but if real life only worked that way.

Another example that raises the question of whether we should do something just because we have the ability to make it happen is brought up in *Once, on a Blue Moon*. Research into bacteriology and virology has allowed us to cure diseases that once ravaged the world. Diseases like polio, tuberculosis, and 'blood poisoning' (blood-borne infections that can lead to infections affecting every organ system, and eventually death) that once struck terror into the hearts of millions are so treatable that most of us never think twice about them anymore.

Unfortunately, the knowledge we have gained in learning how to fight these organisms is a double-edged sword [31]. History is replete with examples of biological warfare, and as our knowledge of bacteriology and virology have advanced, the potential for us to produce more effective weapons has grown as well. But just because we can make more dangerous organisms, does not mean that it is not a very bad idea. No matter how much we think we can contain and control what we produce, there are always variables that will be unaccounted for, variables that could lead to disaster. It's an idea that Hollywood is very fond of, an idea that makes for great drama, but an idea that none of us want to see outside of the movie theater.

Nanomedicine
Stories:
The Last Clone
Hiding from Nobel
A Time to Every Purpose
If He Only Had a Brain
Nanomedicine, as described by Robert A. Freites, Jr., author of the book entitled Nanomedicine, is the "monitoring, repair, construction and control of human biological systems at the molecular level, using engineered nanodevices and nanostructures" [26]. The term *nano* (derived from the Greek work *nanos*, meaning dwarf) is used because these devices are so small they have to be measured in nanometers—one nanometer is 10^{-9} m, thousands of times smaller than the width of a single human hair. We're talking incredibly tiny here.

Like many of my connections with science fiction, it was *Star Trek* that introduced me to the idea of nanomedicine. Although the premise was formulated by much greater minds than screenplay writers, it was the Borg, an evil empire that assimilated anyone they ran into, that sparked my interest in the field. The Borg brought their captives into 'The Collective', by injecting them with nanites—microscopic robots that would enter the cells of the host and transform their prey into part man, part machine.

However, in spite of the usual scenario of science fiction—that all new technology is abused for evil purposes—nanotechnology holds the promise of generating the greatest advances we have ever seen in medicine; perhaps even a cure for aging. Nanomedicine was first mentioned in 1959 by physicist Richard Feynman in a talk entitled *There's Plenty of Room at the Bottom* [25]. His theory was that big tools could be used to make small tools, which in turn could make smaller tools, etc., down to a microscopic scale.

In 1981, an MIT graduate student named K. Eric Drexler suggested that nanodevices might be constructed from biological parts, which could be designed to inspect and repair human cells. His theory was published in 'Smithsonian' magazine in 1982, marking the introduction of nanomedicine to the public (and, no doubt, to the writers of *Star Trek: The Next Generation*), but a more technical version of his work was subsequently rejected by the *Journal of the American Medical Association* as nothing more than "science fiction" [24]. As is usually the case, an increasing body of scientific evidence is gradually swaying the nay-sayers, and science fiction is once again morphing into basic science as nanomedicine gains acceptance as a promising new technology. Freites, one of its greatest proponents, proposed a design for a nanomedicine robot called a respirocyte in the mid 1990s. He described this artificial red blood cell as "a spherical nanobot about the size of a bacterium," and 200 times more efficient at carrying oxygen than a human red blood cell.

Medical research into the use of nanomolecules is advancing on several fronts, including the fabrication of new drugs and drug delivery systems far more effective than any known before. Scientists are designing new types of drugs that can seek out and destroy their targets, harmful invaders like bacteria or cancer cells. Another approach is to engineer new types of delivery systems for drugs we already have; nanomolecules carrying these drugs can be injected into the bloodstream and programmed to circulate throughout the body until they find their target —the location of a tumor, for example. They then release their drugs directly at the site of the tumor, so that much smaller doses of those drugs are necessary, and exposure to healthy tissue elsewhere in the body is limited. The end result is to increase the efficacy of drug treatment while vastly decreasing the harmful side effects.

The possibilities are endless. Though nanomedicine is still in its infancy, we are witnessing the birth of a revolutionary technology that will forever change medical care. Freitas speculates that if we can program nanobots to enter the human body, seek out and repair all malfunctioning cells, the little creatures could, in theory, reverse the damaging effects of cellular aging. Nanobots could one day be the true Fountain of Youth, an idea explored in *The Last Clone*.

The Ethics of Reforming Health Care Delivery

Stories:

Questioning the Tree

If He Only Had a Brain

In an ideal world, everyone would have free access to the best health care science can provide. In the early days of modern medicine, this was certainly in the realm of possibility. Cutting edge science consisted of treatments such as cupping (placing warmed cups against the skin and letting them cool to draw blood to the surface of the skin), leeches (which are still used today, believe it or not, but in very limited, specialized circumstances), and blood-letting (draining the body of blood, which rarely proved to be of tremendous benefit—big surprise there). When this is your big three, it's not too hard to keep the costs down.

In the 1960s there was little to do for a stroke other than physical therapy. Today we have an incredible array of expensive diagnostic tools to root out the cause of the stroke, costly therapies to treat and sometimes reverse the effects of the stroke, and sophisticated equipment to help people recover (computers, robotic devices, technologically advanced bracing, etc.) And with all this, the cost of caring for someone with a stroke has risen dramatically.

Numerous other examples parallel this phenomenon in every branch of medicine. The end result is that medical care now offers a level of care that was unimaginable (except in the minds of fiction writers, of course) only a half-century ago, but at a cost that would have been inconceivable only a half-century ago as well.

As the cost of the latest medical technology rises astronomically, we can no longer afford to provide everything to everybody any more than we could solve the world's transportation problems by providing everyone with a Rolls Royce when they turn 16. As we struggle to find a happy medium, a system where everyone can get at least a basic, high-quality level of care at an afford-able price, the ethics of the various proposals paint the political landscape and become fodder for those of us who like to speculate in fiction.

The Last Clone deals with one aspect of this, when an extraordinarily ex-pensive method of human cloning is developed to extend the lives of only the wealthiest few in society. And *Questioning the Tree* takes a look at one of the potential fallout issues of trying to decrease the cost of medical care by mechanizing and over-regulating a health care system, relegating health care delivery to computers rather than physicians.

We live in an age of computers, which is both wonderful and maddening at the same time. It allows us to do so much, so quickly and efficiently, but makes us more helpless to cope when the computers are down or when a situ-ation falls outside of the norm and requires special attention.

Computerization has revolutionized medical care over the past several de-cades. Computer-guided imaging such as CAT scans can give us images in minutes that allow us to see into the body, to diagnose and treat problems doctors could only guess about not so long ago. Monitoring devices give us continuous graphical readouts of blood pressure, pulse, temperature, blood oxygen levels and more; data that help us rapidly identify problems before it is too late. The examples go on and on, and we all take them for granted.

One of the slower areas to adapt has been computerized health records. Ironically, this may be the least technically difficult area in medicine to computerize, but the difficulty lies in capturing the nuance of an individualized physical exam or the assessment of a physician trained in medical deductive reasoning. It has proven difficult to come up with a standardized system that doesn't make every note look the same, a system where meaningful data can be recorded in an efficient manner, yet still allow each physician to document subtle changes they find on a patient's exam. We are making strides, and most hospitals have adopted electronic records systems that allow rapid access to a wealth of information about each patient, though some areas, such as physician charting, is being incorporated more slowly.

Unfortunately, the ability to computerize and standardize records has resulted in a false confidence that medical care can be improved by having all notes standardized and monitored by an oversight group. We're gravitating toward a system where reimbursement for services is based on accurately filling in boxes on an electronic record; where medical professionals must order certain tests and procedures based on government protocol; where each medical decision made must be weighed against the risk of a law suit. At what point does all this regulation become counterproductive? When we remove the physician's ability to allow critical thinking to guide patient care and replace it with computerized protocols in an attempt to standardize care and avoid errors, are we making that care better or worse?

Questioning the Tree offers one opinion on this very complex issue, an issue that is reshaping the lives of every physician in America, and whether they know it or not, every patient as well; in other words—all of us.

References and Suggested Readings

Robotics—Artificial Intelligence
1. Harris, Tom (2013) **How Robots Work. How Stuff Works.** Accessed Oct 16, 2013. Retrieved from: http://science.howstuffworks.com/robot6.htm
2. Kelly, Kevin (2012) Wired. Better Than Human: **Why Robots Will—And Must—Take Our Jobs.** Retrieved from: www.wired.com/gadgetlab/2012/12/ff-robots-will-take-our-jobs/all/

Robotics—Telerobotic Surgery
3. Anvari, Mehran;McKinley, Craig;Stein, Harvey (2005) **Establishment of the World's First Telerobotic Remote Surgical Service—For Provision of Ad-**

vanced Laparoscopic Surgery in a Rural Community. Annals of Surgery 241(3): 460–464

Robotic Exoskeletal Devices

4. Esquenazi, Alberto; Talaty, Mukul; Packel, Andrew; Saulino, Michael (2012) **The ReWalk Powered Exoskeleton to Restore Ambulatory Function to Individuals with Thoracic-Level Motor-Complete Spinal Cord Injury**. American Journal of Physical Medicine & Rehabilitation 91(11): 911–921
5. Chen, Brian (2012) **New Breed of Robotics Aim to Help People Walk Again**. New York Times. Retrieved from www.nytimes.com

Bionics

6. Zrenner, Eberhart; Bartz-Schmidt, Karl Ulrich; Benav, Heval; Besch, Dorothea; Bruckmann, Anna; Gabel, Veit-Peter; Gekeler, Florian; Greppmaier, Udo; Harscher, Alex; Kibbel, Steffen; Koch, Johannes; Kusnyerik, Akos; Peters, Tobias; Stingl, Katarina; Sachs, Helmut; Stett, Alfred; Szurman, Peter; Wilhelm, Barbara; Wilke, Robert (2011) **Subretinal electronic chips allow blind patients to read letters and combine them to words**. Proceedings of the Royal Society B 278(1711): 1489–1497
7. Chow, Alan Y.; Chow, Vincent Y.; Packo, Kirk H.; Pollack, John S.; Peyman, Gholam A.; Schuchard, Ronald (2004) **The artificial silicon retina microchip for the treatment of vision loss from retinitis pigmentosa**. JAMA Ophthalmology 122: 460–469
8. Graham-Rowe, Duncan (Mar. 2011) **A Bionic Eye Comes to Market**. MIT Technology Review. Retrieved from: www.technologyreview.com
9. Bourzac, Katherine (Feb. 2013) **Bionic Eye Implant Approved for U.S. Patients**. MIT Technology Review. Retrieved from: www.technologyreview.com
10. Chase, Victor (Nov. 2000) **First Bionic Eyes**. MIT Technology Review. Retrieved from: www.technologyreview.com
11. NIDCD Fact Sheet: **Cochlear Implants**. NIH Publication No. 11-4798. Updated March 2011. Retrieved from: www.nidcd.nih.gov/health/hearing/pages/coch.aspx
12. Cullen, D. Kacy; Smith, Douglas H. (2013) **How Artificial Arms Could Connect to the Nervous System**. Scientific American, Jan. 14, p. 52–57

Brain–Computer Interface

13. Nicolelis, Miguel A.L. (2012) **The New Neuroscience of Connecting Brains with Machines—and How It Will Change Our Lives**. St. Martin's Griffin, New York
14. Lebedev, Mikhail A.; Nicolelis, Miguel A.L. (2006) **Brain–machine interfaces: past, present and future**. Trends in Neurosciences 29(9): 536–546
15. Hochberg, Leigh R.; Serruya, Mijail D.; Friehs, Gerhard M.; Mukand, Jon A.; Saleh, Maryam; Caplan, Abraham H.; Branner, Almut; Chen, David; Penn, Richard D.; Donoghue, John P. (2006) **Neuronal ensemble control of prosthetic devices by a human with tetraplegia**. Nature 442: 164–171

16. Hochberg, Leigh R.; Donoghue, John P. (2006) **Sensors for brain-computer interfaces**. Engineering in Medicine and Biology Magazine, IEEE 25(5): 32–38

Nanomedicine

17. Youfan, Hu; Yan, Zhang; Chen, Xu; Long, Lin; Snyder, Robert L.; Zhong, Lin Wang (2011) **Self-Powered System with Wireless Data Transmission**. Nano Letters 11: 2572–2577

18. Freitas, Robert (1999) **Nanomedicine**, Vol. 1: Basic Capabilities. Georgetown, Texas: Landes Bioscience

19. Behan, Nihall (2013) **Nanomedicine and Drug Delivery at the University of Limerick**. Accessed Oct 16, 2013. Retrieved from: www3.ul.ie/~childsp/elements/issue4/behan.htm

20. **Nanotechnology Newsletter.** GlobalSpec.com: http://www.globalspec.com/newsletter/pub/15/nanotechnology

21 Burgess, Rob (2012) **Understanding Nanomedicine**. Pan Stanford Publishing Pte. Ltd., Singapore

22. Ventola, C. Lee (2012) **The Nanomedicine RevolutionPart 2: Current and Future Clinical Applications**. Pharmacy and Therapeutics 37(10): 582–591

23. Ellis-Behnke, Rutledge G.; Liang, Yu-Xiang; You, Si-Wei; Tay, David K. C.; Zhang, Shuguang; So, Kwok-Fai; Schneider, Gerald E. (2006) **Nano neuro knitting: Peptide nanofiber scaffold for brain repair and axon regeneration with functional return of vision**. Proceedings of the National Academy of Science U.S.A. 103(13): 5054–5059

24. Freitas, Robert (July/Aug. 2000) **The Sciences**. The New York Academy of Sciences. Retrieved from: http://www.foresight.org/Nanomedicine/SayAh/

25. Feynman, Richard (Dec. 1959) **Plenty of Room at the Bottom**. Retrieved from: http://www.its.caltech.edu/~feynman/plenty.html

26. Foresight Nanotech Institute. **Nanomedicine**. Retrieved from: www.foresight.org/Nanomedicine/

Ethics of Cloning

27. Tierney, John (Nov. 20, 2007) **Are Scientists Playing God? It Depends on Your Religion**. New York Times. Retrieved from: http://www.nytimes.com/2007/11/20/science/20tier.html?pagewanted=all&_r=1&

28. **Report of the Council on Ethical and Judicial Affairs of the American Medical Association.** June 1999. Retrieved from: http://www.ama-assn.org/resources/doc/ethics/report98.pdf

29. **The Cloning Technique Used to Clone Dolly.** Retrieved from: http://www.bioinformatics.nl/webportal/background/dollyinfo.html

Ethics of Biological Warfare

30. Dire, Daniel J. (Chief Editor: Darling, Robert G.) Medscape. **Biological Warfare Agents**. Retrieved from: http://emedicine.medscape.com/article/829613-overview

31. Reyes, Daniel (2003) **The Ethics of Biowarfare**. ActionBioscience.org. Retrieved from: http://www.actionbioscience.org/newfrontiers/reyes.html